An Introduction to the Invertebrates

So much has to be crammed into today's biology courses that basic information on animal groups and their evolutionary origins is often left out. This is particularly true for the invertebrates. The second edition of Janet Moore's *An Introduction to the Invertebrates* fills this gap by providing a short updated guide to the invertebrate phyla, looking at their diverse forms, functions and evolutionary relationships.

This book first introduces evolution and modern methods of tracing it, then considers the distinctive body plan of each invertebrate phylum, showing what has evolved, how the animals live, and how they develop. Boxes introduce physiological mechanisms and development. The final chapter explains uses of molecular evidence and presents an up-to-date view of evolutionary history, giving a more certain definition of the relationships between invertebrates. This user-friendly and well-illustrated introduction will be invaluable for all those studying invertebrates.

Janet Moore is former Director of Studies in Biological Sciences at New Hall, Cambridge, where she is now an Emeritus Fellow. Her research career has focused on land and freshwater nemertines.

An Introduction to the Invertebrates

Janet Moore
New Hall, Cambridge

Illustrations by Raith Overhill

Second Edition

CAMBRIDGE
UNIVERSITY PRESS

CAMBRIDGE UNIVERSITY PRESS

Cambridge, New York, Melbourne, Madrid, Cape Town,
Singapore, São Paulo

Cambridge University Press
The Edinburgh Building, Cambridge CB2 2RU, UK

Published in the United States of America by
Cambridge University Press, New York

www.cambridge.org
Information on this title: www.cambridge.org/9780521857369

First published 2001
Reprinted 2003
Second edition published 2006

Printed in the United Kingdom at the University Press, Cambridge

A Catalogue record for this publication is available from the British Library

ISBN-13 978-0-521-85736-9 hardback
ISBN-10 0-521-85736-8 hardback

ISBN-13 978-0-521-67406-5 paperback
ISBN-10 0-521-67406-9 paperback

Contents

Boxes

first two, are more general, discussing first animal development and then the evolutionary relationships revealed by fossils and by molecules. The text is illustrated throughout by specially prepared figures, many of them new for this edition. Many of these are diagrams of generalised examples and therefore have no scales indicated on the figures.

This book aims to discern and communicate overall patterns because the human mind cannot comprehend a scatter of unrelated facts. Classification is kept to a minimum, with just enough subdivision of phyla to provide group names needed for discussion; in general the aim is to use no more terminology than is necessary for communication. Generalisations are made that, on further enquiry, will require qualification, but this does not obviate the need for a framework of generalisations.

The book arises from many years of teaching undergraduates from many colleges (including my own, New Hall) at Cambridge University. It is written in the hope that others will catch and share an enthusiasm for invertebrate animals.

Acknowledgements

For help in preparing this second edition of *An Introduction to the Invertebrates* it is a pleasure to thank Professor Michael Akam, both for reading and improving the last two chapters and for continued hospitality in the Museum Molecular Laboratory. I am grateful also to Dr Ronald Jenner for information and insight, to Dr Lloyd Peck for the sight of Antarctic animals and information about their physiology, and to my son Dr Peter Moore for spotting errors in the first edition. At Cambridge University Press I should like to thank Dr Ward Cooper for his help and welcome towards a second edition, his successor Dr Dominic Lewis for his help in the later stages, Hugh Brazier for copy-editing, and Dawn Preston for seeing the book through production. Once again I have been privileged to have all the illustrations (many from the first edition but a number of new ones) drawn and prepared by Mr Raith Overhill, to whom I am very grateful indeed.

It is a pleasure once again to thank those who read and improved the first edition of this book: firstly Dr Norman Moore and Dr Peggy Varley, who have read it all, and Dr Barbara Dainton, Dr William Foster and Dr Martin Wells, who have each read and helped with a number of chapters. For reading particular chapters I thank Professor Michael Akam, Professor Ray Gibson, Dr Liz Hide, Dr Hugh Jones, Dr Vicky McMillan, Dr Pamela Roe, Dr Max Telford, Professor Pat Willmer and unknown referees. I am grateful to them all for corrections and helpful suggestions; of course, remaining infelicities and mistakes are my sole responsibility. I should like to thank Dr Tracey Sanderson of Cambridge University Press for her help, guidance and encouragement throughout the writing of the first edition of this book. I should also like to thank Professor Ray Gibson for 30 years' happy collaboration in research on terrestrial and freshwater nemertines and Professor Pat Willmer for educating me even more than most of my past students did, and I extend that thanks to all my students. I am grateful to the late Dame Rosemary Murray and New Hall for many delightful teaching opportunities, and to Professors Gabriel Horn and Malcolm Burrows for hospitality in the Cambridge Zoology Department.

Finally I want to record my gratitude to the late Professor Carl Pantin, my research supervisor, and Dr Barbara Dainton, my undergraduate supervisor, for teaching me and arousing my interest in Zoology, and to my husband Dr Norman Moore for sustaining and sharing that interest.

Illustration acknowledgements

All the illustrations were prepared by Raith Overhill, to whom I am most grateful. They were based either on original material or on illustrations in previously published works. While none of these has been reproduced directly, every effort was made to gain permission to adapt illustrations from copyright holders, who are gratefully acknowledged as follows:

Blackwell Science. Barnes, R. S. K. (ed.) (1998) *The Diversity of Living Organisms*: Fig. 5.018. Barnes, R. S. K., Calow, P. and Olive, P. J. W. (1988) *The Invertebrates: a New Synthesis*: Figs. 3.51, 3.59, 4.9, 4.14, 4.24, 4.25, 4.26, 4.29, 4.35, 4.41, 4.42, 4.45, 4.64, 6.9, 7.1, 7.17, 7.20, 8.12, 8.31, 8.32, 8.40, 8.41, 8.48, 8.49 and 16.20. Barnes, R. S. K., Calow, P. and Olive, P. J. W. (2001) *The Invertebrates: a New Synthesis*, 3rd edn.: Figs. 4.65 and 5.21. Danielsson, D. (1892) *Norwegian North Atlantic Expedition (1876−1878). Rep. Zool.* **21**, 1−28: Fig. 7.15. Marion, M. A. F. (1886) *Archives de Zoologie Expérimentale et Générale*. (2) **4**, 304−326: Fig. 7.6.

Buchsbaum, R. (1938) *Animals Without Backbones*, published by University of Chicago Press: figures on pages 79, 301 and 310.

Cambridge University Press. Borrodaile, L. A., Eastham, L. E. S., Potts, F. A. and Saunders, J. T. (1958) *The Invertebrata*: Figs. 109B, 259 (left-hand drawing), 304, 408, 422, 444 and 486. Denton, E. J. and Gilpin-Brown, J. B. (1966) On the buoyancy of the pearly *Nautilus*, *Journal of the Marine Biological Association* **46**, 723−759: Figs. 1 and 4. Trueman, E. R. (1975) *The Locomotion of Soft-bodied Animals*: Figs. 2.17 and 3.11. Young, D. (1989) *Nerve Cells and Animal Behaviour*: Figs. 2.2, 2.3 and 2.6b.

Company of Biologists Ltd. Gray, J. and Lissmann, H. W. (1938) *Journal of Experimental Biology* **15**, 506−517: Fig. 1. Gray, J. and Lissmann, H. W. (1964) *Journal of Experimental Biology* **41**, 135−154: Fig. 1. Weis-Fogh, T. (1973) *Journal of Experimental Biology* **59**, 169−230: Fig. 21A.

English University Press. Chapman, R. F. (1998) *The Insects: Structure and Function*: Figs. 5.7, 8.1C, 8.1D, 9.1, 16.2, 16.6B, 16.17 and 22.1A.

Garland. Alberts, B. *et al.* (1983) *Molecular Biology of the Cell*: Figs. 14.9 and 15.3.

Gibson, Ray. (1972) *Nemerteans*, published by Hutchinson: Figs. 6, 8A, 8B and 13H.

Kluwer Academic Publishers. Wells, M. J. (1978) *Octopus: Physiology and Behaviour of an Advanced Invertebrate*, Chapter 2, Anatomy: Fig. 2.1 (p. 13). Wigglesworth, V. B. (1965) *Principles of Insect Physiology*, 6th edn, Chapter 9, Respiration: Figs. 228 (p. 322) and 234 (p. 328). With kind permission from Kluwer Academic Publishers.

Oxford University Press. Wolpert, L. *et al.* (1997) *Principles of Development*: Figs. 13.7 (p. 402), 13.8 and 13.9 (p. 403). By permission of Oxford University Press.

Reed Educational and Professional Publishers Ltd. Freeman, W. H. and Bracegirdle, B. (1971) *An Atlas of Invertebrate Structure*: Fig. 50. Reprinted by permission of Heinemann Educational Publishers, a division of Reed Educational and Professional Publishers.

Weidenfeld and Nicolson. Wells, M. J. (1968) *Lower Animals* (published in the World University Library series): Figs. 1.3, 2.3, 2.5, 4.7, 9.3, 10.2, and parts of 4.8, 4.9, 5.5, 6.9, 9.6B and 10.4.

Worth Publishers. Eckert, R. and Randall D. (1978) *Animal Physiology* (W. H. Freeman & Company): Fig. 9.8. Used with permission.

Chapter 1

The process of evolution: natural selection

This book is about invertebrate evolution. Every account of structure and function and the adaptation of an animal to its environment is a description of the results of evolution. Not only the intricate design but also the vast diversity of animals has been achieved by descent with modification due to the action of natural selection. A process so fundamental needs to be introduced at the very beginning of the book. As the different phyla are presented, general discussion of some other topics will become necessary (and will be inserted as 'Boxes'), but evolution cannot wait.

1.1   What was Darwin's theory of natural selection?

Our understanding of evolution dates from the publication in 1859 of Charles Darwin's great book *The Origin of Species by Means of Natural Selection, or the Preservation of Favoured Races in the Struggle for Life.* Before that time, explanation of all the details of animal design in terms of a divine Creator was widely accepted, though perhaps the extraordinary variety of life (e.g. what has been termed 'the Almighty's inordinate fondness for beetles') was harder to explain. From very early times a few writers had postulated evolutionary theories, suggesting that different species might not all have been separately created, and further that complicated forms of life could have arisen from simple antecedents by descent with modification. This however was mere speculation in the absence of support from a large array of ordered facts. What Darwin gave us was a mass of careful observations, many gathered while he was Naturalist on the voyage of HMS *Beagle*, from which he formulated a theoretical framework showing that evolution could have occurred by what he called 'Natural Selection'. That the time was ripe for such a theory is shown by the simultaneous conclusions of Alfred Russel Wallace from his work in Indonesia. The cooperation of Darwin and Wallace without any competition for priority is an encouraging example of decency transcending competition.

Darwin's argument was as follows:

1. Living things tend to multiply. There are more offspring than parents and, if unchecked, their numbers would increase in geometrical ratio.
2. The progeny cannot all survive, because resources (food, space, etc.) are insufficient.

Therefore there will be competition for survival, a 'struggle for existence' between individuals of the same species.

3. Living things vary; the progeny are not all identical and some will be better equipped for survival than others.

Therefore 'favourable variations would tend to be preserved, and unfavourable ones to be destroyed. The result of this would be the formation of new species' (*The Autobiography of Charles Darwin*, ed. Norah Barlow, Collins 1958, p. 120).

To describe this process of natural selection Herbert Spencer used the phrase 'survival of the fittest'. The phrase needs to be qualified if misunderstanding is to be avoided: firstly, it is not mere survival but differential reproduction that is required and, secondly, 'fittest' does not refer to general health and strength but to some precise advantage in particular circumstances in a particular environment. Adaptation consists in the perpetuation of such an advantage down the generations.

Here at once was Darwin's greatest difficulty. For natural selection to work, advantageous changes had to be inherited. In Darwin's time heredity was assumed to involve the blending of the features of the two parents, and Darwin was much worried by the criticism (from an engineer, Fleeming Jenkin) that any system of blending inheritance would remove the advantage in a few generations. The solution was at hand, but never known to Darwin. Gregor Mendel had already shown that heredity was particulate, but his work was not publicised until 1900.

1.2   What was Mendel's theory of heredity?

Mendel's 'atomic theory' of heredity was based on his experiments on crossbreeding garden peas. He deduced that hereditary factors are constant units, handed down unchanged from parent to offspring, and that these units occur as 'allelomorphic pairs', the two members of each pair representing two contrasting characters. At sexual reproduction when gametes (spermatozoa and ova) are formed, only one factor of each pair can enter a single gamete. When gametes fuse to form a 'zygote' the factors, one from each parent, are combined. One factor in a pair may be 'dominant' over the other, which is called

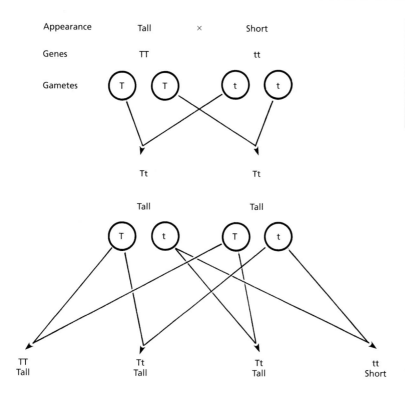

Fig. 1.1 The Mendelian ratio: showing that, if 'T' is tall and 't' is short, the first generation will all appear tall and that (after self-pollinating) the second generation will have the ratio of three tall to one short.

the 'recessive' and has no apparent effect on the organism but is maintained when it reproduces (Figure 1.1). The organism contains a very large number of such pairs (some 50 000 pairs in humans) most of which segregate and recombine independently at every sexual reproduction. Mendel's analysis explained both the basic resemblance between parents and offspring and the introduction of variation between them.

1.3   What is the cellular basis of heredity?

Early in the twentieth century, T. H. Morgan's studies of cell structure identified Mendel's factors as 'genes' borne on the elongated bodies, 'chromosomes', contained in the nucleus of almost every cell in the body (see Chapter 15, where the contributions of studies of the fruit fly *Drosophila melanogaster* are discussed). All organisms develop from the division of cells which previously formed part of one or (where reproduction is sexual) two parent organisms. August Weismann first recognised that the 'germ-plasm' that gives rise to gametes is distinct from the rest of the body, the 'soma'. Somatic cells divide by 'mitosis', the longitudinal splitting of each chromosome with self-replication of each gene so that each half chromosome has exactly the same genes as its parent (Figure 1.2a): all the somatic cells in an individual are genetically identical. Gamete-forming cells first multiply by the

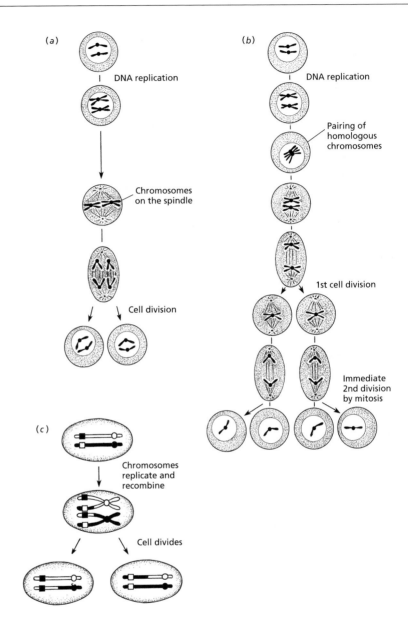

Fig. 1.2 Cell division: (a) mitosis, (b) meiosis, (c) crossing over. Squares represent genes, circles represent points of junction.

(a)

DNA replication

Chromosomes on the spindle

Cell division

(b)

DNA replication

Pairing of homologous chromosomes

1st cell division

Immediate 2nd division by mitosis

(c)

Chromosomes replicate and recombine

Cell divides

same process of mitosis but then divide by 'meiosis', a process in which the number of chromosomes is halved and (usually) the two genes in each allelomorphic pair are separated as Mendel postulated (Figure 1.2b). The fusion of gametes combines half the genes of each parent to make a new individual. Gametes are described as 'haploid' since they contain half the number of chromosomes of the 'diploid' zygote and adult individual.

Only the gametes carry genes to the next generation. Changes often occur in the somatic cells, caused by use or disuse or by direct effects of the environment, but such changes cannot be transmitted to the offspring. Jean-Baptiste Lamarck is rather unfairly remembered mainly for his erroneous belief in the inheritance of acquired characters. Lamarckism has been typified by the idea that if giraffes stretched their necks to reach more food their offspring would be born with longer necks. A change to the body such as an elongated neck cannot directly affect subsequent generations: they can be changed only by the selection of individuals with genes promoting the growth of long necks. The 'phenotype', that is the organism defined by the characters made manifest, must be distinguished from the 'genotype' or genetic constitution, which alone can transmit changes to the offspring.

Note that the word 'develop' was originally used to describe two different consequences of gene action: the sequence of changes in an individual as the egg gives rise to the adult form, called 'ontogeny', and (on an enormously greater time scale) the process of evolutionary change, called 'phylogeny'. We now reserve the term 'development' for ontogeny.

1.4   What is the origin of genetic variation?

Genes provide both the continuity and the differences between parents and offspring. The differences ('variations') are caused as follows:

1. Combination of half the genes from each parent.
2. Reassortment of the genes inherited from each parent. Genes borne along the same chromosome tend to be inherited together (they are said to be 'linked') but during meiosis there is normally some 'crossing over', or exchange of pieces of the split chromosomes (Figure 1.2c).
3. The presence of a gene does not guarantee the appearance of the character with which it is associated, because gene effects may depend on the action of other genes present. The simplest example is dominance within an allelomorphic pair, but other genes may promote, suppress or alter the effect of a gene. A character may be the product of many different genes acting together, and one gene may affect many characters; for example, genes acting early in development may transform the effects of other genes acting later. It is a dangerous oversimplification to equate a character with the gene that in part governs it. Mendel has been mistakenly described as 'lucky' because his choice of the peas gave a simple picture: in fact he spent a very long time experimenting to find suitable material.
4. Mutations occur. These may be chromosome changes, or more frequently errors in gene copying as cells divide. Sudden change in a phenotype due to mutation is rarely advantageous,

as large changes tend to be lethal, but small changes may accumulate in the genotype, undetected until some change in circumstances gives them a selective advantage.

Clearly, mutation is the only one of these causes of variation that operates in asexual reproduction, where otherwise parent and offspring are genetically identical.

With the mechanism producing heritable variations understood, the picture of evolution caused by natural selection acting on random variations became firmly established. Due mainly to R. A. Fisher, the emphasis fell not on the sudden change in form of an individual but on the spread of that variation through a population. The study of natural selection at work became a matter of statistics rather than qualitative descriptions. The synthesis of Mendelian genetics and natural selection was called Neo-Darwinism or 'The Evolutionary Synthesis', and by the 1930s it was widely accepted. It became the unifying principle underlying all branches of biology.

1.5   What is the nature of genes?

The work of James Watson and Francis Crick and others revealed in 1953 that DNA, in the form of a double helix, is the genetic material in the chromosomes. It replicates when the cell nucleus divides, and it can be transcribed to make RNA, giving a message that in turn can be translated into assembly of amino acids to make proteins. Genes and their action can now be studied at the molecular level, which has led to an enormous increase in understanding and opportunities for manipulation. There are also new problems: for example, some at least of the changes at the molecular level may not be due to natural selection. Could there be evolutionary change due simply to chance?

1.6   What is the role of chance in evolution?

This has frequently been misunderstood by critics, some of whom regard the whole process as a combination of lucky accidents. They fail to distinguish the two stages involved. Variations arise by chance, as mutations and gene recombination occur at random. What is not at all a matter of chance is the operation of natural selection, which acts on these random variations to produce adaptation. Certainly change in gene frequencies may partly be due to chance; for example, long ago Sewall Wright pointed out that a small isolated population would contain only a few of that species' genes and therefore these genes would become over-represented in the population. This process is called 'genetic drift' and cannot itself cause adaptation. It is still not clear whether 'neutral' (i.e. unselected) molecular evolution is important.

1.7 At what level does natural selection act?

Our understanding of evolution continues to evolve. Natural selection was at one time assumed to act for the good of the species or the group, until both experiment and theory showed that natural selection acts on individuals and cannot be shown to act on any larger entity. This at once produced new problems: a prominent puzzle is to find the advantage of sexual reproduction. Clearly this is slow and complicated compared with simple asexual multiplication, but of enormous benefit to the species because it introduces so much variation. How can this advantage apply at the individual level, by an organism being unlike its parents? The problem is unresolved, but explanations focus on the masking of harmful mutations or on the so-called 'Red Queen' effect: the need to run as fast as possible simply to keep up. Host and parasite, for example, engage in a continual 'arms race': host offspring differing from their parents have more chance of avoiding their parasites, and parasites chemically different from their parents may evade the host's defences.

1.7.1 The unit of selection

Natural selection acts on the individual, but the effect of this action is the passing on of one set of genes rather than another. It is the relative frequency of genes that changes down the generations. As has been cogently argued by Richard Dawkins (in *The Selfish Gene*), the individual's body is the vehicle for the genes, which are the replicators; individual bodies are the genes' way of preserving the genes unaltered. Arguments about whether the individual or the gene is the 'unit of selection' are unprofitable: the important thing is to remember the role of each. Natural selection acts on the phenotype, not directly on the genotype. The danger of equating genes and characters must not be forgotten. Further, no gene has a fixed selective value: its effect will depend upon other genes present. Genes are now known to change surprisingly little during evolution; what changes is the regulation and expression of those genes, as will be explained and illustrated in Chapter 20.

The above brief outline may serve to introduce invertebrate evolution, but further reading (see the end of this book) is strongly recommended, to supply evidence for the above assertions and fuller discussion of these and many more facts and ideas.

1.8 What in general does evolution produce?

1.8.1 Diversity

Diversity is the product of evolution. The very long evolutionary history of invertebrates has allowed an abundance of diversity: the very name 'invertebrates' is revealing: they can only be united

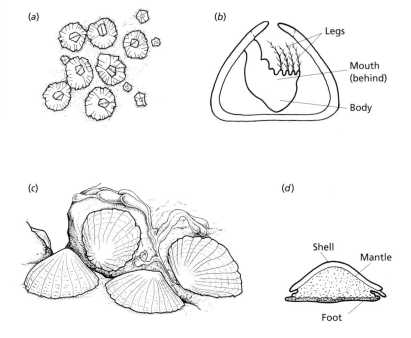

Fig. 1.3 Barnacles and limpets on the seashore: (a) acorn barnacles; (b) vertical section through a barnacle (diagrammatic); (c) common limpets; (d) vertical section through a limpet (diagrammatic).

as 'animals other than vertebrates'. Selection pressure caused divergence among the earliest multicellular forms, and certain body plans became successfully established, each offering particular opportunities and constraints for evolution, as this book will show. Animals sharing the same body plan are united in a 'phylum' (plural 'phyla'). Within each phylum there are usually well-defined classes with characteristics that fit the animals for some particular environment or way of life, and within each class the original body form will have become modified as they exploit different habitats. This diversification from a single ancestral form is known as 'adaptive radiation'.

At the same time, natural selection does not only produce novelty: it also maintains and stabilises a successful structure and way of life. Nor does it only produce divergence: animals of very different ancestry may become very similar through adaptation when they solve the same problems or live in the same environment. Convergence is a widespread, often undervalued cause of resemblances, often baffling the attempt to classify animals by tracing their evolutionary history. For example, on an exposed shore the barnacles and limpets are superficially similar, being attached to rocks and covered by shells that protect them from desiccation and predatory birds (Figure 1.3). When the tide is up, the barnacle is seen to be a crustacean, shrimp-like but attached by its head and kicking its food into its mouth with its legs, while the limpet is a mollusc with a muscular foot like a snail, moving off in search of food. This is a crude example: one has only to think of the mimicry in butterflies from different families to realise the fine degree of convergence that can be produced by natural selection, superimposed upon its primary divergent effect.

Natural selection defies man-made categories: for example, in defining Platyhelminthes we state that the mouth is the only opening to the gut, yet one parasitic species has not just one anus but two. Animals are opportunists. Our categories especially meet trouble when we try to define a species, because we are trying to put firm boundaries on an evolutionary continuum. If species were incapable of changing, evolution could not have occurred.

1.8.2 Complexity

As well as diversity, complexity is a product of evolution. Primitively, multicellular forms were not very complex (note that 'primitive' means 'most like the ancestral form', not 'simplest'). What we call the 'higher' animals, those more recently produced, are on the whole very much more elaborate than their ancestors. The evolutionary pattern is clear, but must at once be modified by what we know of the evolutionary process. Complexity is not an end in itself: it will evolve where it has selective value, but not otherwise. Many simple forms survive today: one has only to look at sponges, animals extremely successful in that they are very numerous and widely distributed in the sea, yet remarkably simple in structure. A sponge is the best way of being a sponge, and natural selection has not in millions of years produced much alteration in their form. Cnidaria (anemones, corals, jellyfish, hydroids) again are simple in structure but remarkably numerous in the sea (unlike sponges they were also able to evolve great morphological diversity, as will be shown). In the more elaborate phyla the simplest animals are not necessarily the most primitive, as is clearly illustrated within the Platyhelminthes. Simplicity may be a secondary product of adaptation, and we cannot assume that a simple animal is primitive.

1.8.3 Not progress

Evolution is not directed from the outside and there is no inner directing force. The criterion for survival is immediate selective advantage, not any long-term evolutionary aim. We are being anthropocentric when we misapply the idea of progress to the evolutionary process. We like to think of all evolution leading up to humans at the apex of the evolutionary tree. This is a false picture.

1.8.4 Efficiency

This term needs careful definition. While natural selection should tend to maximise efficiency, that does not always mean maximum physical efficiency: biological efficiency can be different. For example, that a constant high body temperature enables the body's enzymes to work at maximum speed may be physically most efficient, yet it may be more advantageous to an animal to let the body temperature fluctuate, allowing the economy of cold inactive periods.

Natural selection does not necessarily generate our own idea of a perfect product. 'Success', another anthropocentric approach, cannot be defined in terms of complexity or position in an evolutionary tree, but rather in terms of survival, abundance and perhaps also diversity.

As we study the invertebrate phyla that are the products of the evolutionary process, we can safely ask the question 'Why?' (as in 'Why is this animal so constructed?'). This is because we know that what a biologist means by such a question is 'How has such a structure conferred selective advantage?' Long ago when I was a student a professor said to us 'When anyone asks me the question "Why?" I refer him to a theologian.' I now think he was wrong — on both counts, because theologians cannot answer such questions and biologists can make the attempt.

Chapter 2

The pattern of evolution: methods of investigation

2.1 How should we classify animals?

Classification is essential to any study of animals (the first attempt is attributed to Adam) and is a necessary prelude to tracing the pattern of evolution. Systematic ordering of the products of classification (taxonomy) can be done in various ways, but most usefully it aims to produce a 'natural' classification, i.e. a phylogeny that reveals evolutionary history. We try to put together those animals most closely related by descent, using resemblance as the basis for our classification.

Classification is difficult. A long-ago cartoon in the magazine *Punch* showed a railway porter scratching his head and saying 'Cats is dogs, and rabbits is dogs, but this 'ere tortoise is an insect.' We must sympathise with his dilemma. The difficulty is that resemblance between animals is not an entirely reliable guide to their evolutionary history. What makes animals resemble each other? It may be due either to close common ancestry or to convergence, occurring when animals of different ancestry acquire very similar adaptations because they face the same problems or live in the same environment. For example, any land-living invertebrate will have a skin relatively impermeable to water; this character is no guide to closeness of ancestry, it is due to the general need to avoid desiccation.

The main challenge in biological classification is to distinguish these two causes of resemblance: homology, where there is a common evolutionary origin, and convergence, where similarity emerges from different evolutionary pathways. Convergence is very widespread but difficult to prove, and it is difficult even to define homology precisely, as the final chapter (20) will explain. The great change in recent years is that we have new procedures of classification and new sources of information that may now help us to distinguish homology from convergence and therefore to trace invertebrate phylogeny with greater confidence. We need no longer agree with Libbie Hyman, who in 1959 wrote that when we attempt to relate phyla, 'anything said on these questions lies in the realm of fantasy'. This chapter introduces our new tools.

2.2 How can we use morphology to trace phylogeny?

2.2.1 The traditional method

Animals are studied and compared. Characters indicating resemblance are picked out and assessed, as to whether they are independent of each other and whether the resemblance is likely to be due to convergence. Evidence is drawn from fossils, embryology, geographical distribution and any other available source. This process is subjective, yet it can be invaluable to use the opinion of an experienced systematist who has studied the particular group of animals. An advantage of this method is that careful assessment of characters is inescapable.

2.2.2 Phenetic taxonomy

'Phenetic' means 'as observed in the phenotype', and phenetic taxonomy originally meant 'classification by observed similarity', as in the traditional method. Phenetic taxonomy now frequently has a more restricted meaning, being equated with numerical taxonomy, a method where animals with the greatest number of common characters are put together. This process appears to be objective but there are concealed subjective steps: characters are selected and defined and then assumed to be finite and equivalent units, independent of each other. Assessment of the characters is vital, but the procedure does not ensure that it will be made. There is no attempt to identify convergence, which is assumed to be much less common than resemblance due to common ancestry and therefore likely to be eliminated when large numbers of characters are used. This method fails when convergence is common, or when only small numbers of independent characters are available.

2.2.3 Cladistic analysis

This is also called 'phylogenetic classification'. It groups organisms according to recency of common ancestry, as revealed by the presence of 'shared derived characters', i.e. when animals share the same difference from the primitive condition (Figure 2.1a). For example, a primitive character shared by all insects, such as the possession of an exoskeleton, is not a character helpful for determining the relationships of a cockroach, a bee and a wasp, but the presence of hooks 'marrying' the two wings on each side of a bee and a wasp is a 'derived' ('specialised') character, present in both, which does distinguish them from the cockroach (Figure 2.1b). The nodes (branching points) define differences from the primitive condition

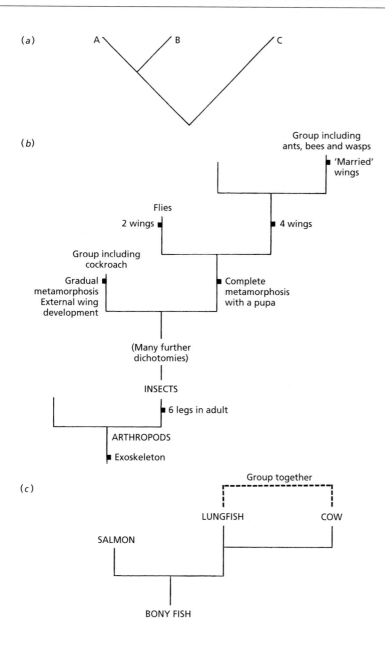

(a)

(b)

(c)

Fig. 2.1 (a) Basic cladogram; (b) the cockroach, the bee and the wasp; (c) a hierarchy of recency of common ancestry.

and these differences will be shared by all the lines beyond that node. These are the shared derived characters (sometimes called 'synapomorphies'). Cladistic analysis aims to identify 'monophyletic groups', i.e. the ancestor and all of its descendants. On the cladogram, such a group will be represented by a node and everything following from it. Time is not represented: the length of a branch has no significance. The sequence of common ancestry is all that can be deduced (Figure 2.1c).

How reliable is such a cladogram? The emphasis on differences, which can be defined more precisely than similarities, and on the need to distinguish between primitive and derived characters, is very helpful. Much information is systematically obtained and made available for comparisons. There are, however, considerable drawbacks to cladistic classification, and now that the procedure is very widely adopted it is important to recognise these drawbacks. They are:

1. Characters are too readily treated as fixed finite units that correspond in different groups, for example 'proboscis present' does not unite a worm and an elephant when 'proboscis' denotes totally different structures. This drawback can be overcome by sufficiently careful assessment of the characters used; the trouble is that cladograms can be drawn without such an assessment. The world is awash with ill-considered cladograms.

2. Rooting the cladogram (i.e. defining the primitive condition) is often very difficult. The root is compared with an 'outgroup', chosen as being related to the group but not part of it. Choice of the outgroup is subjective and difficult. It has reasonably been claimed that whenever outgroup analysis can be applied unambiguously it is not needed, and whenever it is needed it cannot be applied unambiguously.

3. The worst problem is that usually many possible cladograms can be drawn, and to pick the correct one the principle of 'parsimony' is invoked. Parsimony in this context means selection of the cladogram with the smallest number of evolutionary steps. The assumption here is the rash one that resemblance due to close common ancestry is much commoner than resemblance due to convergence; the result is that perception of convergence is minimised. The difficulty of choosing between what may be a large number of equally 'parsimonious' cladograms is a further problem.

Despite these drawbacks, cladistic classification has been widely adopted and has revived interest in tracing animal relationships using morphological evidence.

2.3 How can we use fossils to investigate phylogeny?

Fossils provide morphological evidence about animals that lived long ago.

2.3.1 What are fossils?

They are the remnants of once-living animals, preserved in rocks or (less often) in sediments, amber, ice etc. The term is normally reserved for remains dating back to before the last Ice Age. Hard parts of animals may be preserved with little change in appearance: among invertebrates these include arthropod exoskeletons, the shells of molluscs and brachiopods (another phylum of shelled animals), echinoderm skeletons and jaws and other hard bits from many phyla. Arthropod exoskeletons may be fossilised whole or preserved as thin films of carbon on rocks. Fossilised soft parts have usually been turned to stone, for example by replacement of organic material by minerals from solutions underground.

> **Living fossils**: this is a potentially confusing description of animals that have changed remarkably little over long periods of time. Examples include the brachiopod *Lingula*, found in Cambrian fossils and persisting today.
>
> **Trace fossils** are imprints on the environment of the activity of animals long ago, such as tracks, trails, burrows, coprolites (i.e. fossilised faeces) and impressions on soft substrata which may be made even by animals such as jellyfish.

2.3.2   What can fossils tell us about evolution?

The fossil record can tell us about the structure and way of life of past animals and the sequence in which they appeared: the facts about the time dimension may enable us to root evolutionary trees. Obviously the fossil record is very patchy and very incomplete: animals lacking hard skeletons are less likely to leave fossil evidence. Even where a group has as many fossils as the arthropods, some animals once very common (such as the trilobites) occur in large numbers while rarer forms may not be represented at all. Extinct animals may present us with a baffling mixture of characters. Nevertheless, not only can the fossil record tell us about the morphology of past organisms, it can also with careful interpretation reveal facts about animal mechanics, and also about ecological interactions and the nature of past ecosystems.

The succession of life forms that can be traced is sufficiently reliable for geologists to date rocks by the fossils which they contain (Figure 2.2). As the sequence in which the main groups of animals occurred is determined, new finds continually push the earliest appearance of each group back in time. Direct ancestry, however, is not revealed. While every fossil must have a nearest living relative, we can only very rarely identify it, and in any case the chance of finding a direct ancestor is vanishingly small. Among fossils as among living animals, a supposed 'missing link' between phyla can seldom be authenticated, although it may be disproved. Yet where fossils are plentiful a group of intermediate and possibly transitional forms can sometimes be identified — information that molecules can never provide.

Barnacles on the seashore give us a nice example of information about evolutionary change. From the earliest fossils to modern forms there is a trend of reduction of the number of lateral plates round the body. The rich fossil record enables us to trace this reduction in seven of the eight lineages of barnacles in this family, a remarkable example of parallel evolution. Why should it have occurred? Fewer plates means fewer junctions between plates, and observations

	Ma	(millions of years ago, to the beginning of each period)	
CENOZOIC	65	Tertiary	Pangea drifted apart, birds and mammals spread.
MESOZOIC	140	Cretaceous	Ended at K/T border with extinctions of dinosaurs, ammonites and much else.
	240	Jurassic	Earliest birds and mammals.
	250	Triassic	Appearance of teleost fish, dinosaurs.
PALAEOZOIC	300	Permian	Continents join as Pangea, including Gondwanaland. Climate deteriorated, loss of diversity, extinctions of trilobites, many brachiopods and ammonites. Ended in an ice age.
	360	Carboniferous	Warm seas, food plentiful, great diversity. Forests: trees with seeds. Amphibia, reptiles, giant dragonflies.
	410	Devonian	Old red sandstone, horsetails. Graptolites gone. Age of fish, all kinds, and first tetrapod vertebrates.
	430	Silurian	Making of mountains. First land plants and arthropods.
	500	Ordovician	Continental drift; more food, PO_4, O_2. Trilobites, graptolites, nautiloids, coral reefs. Many Cambrian oddities replaced by familiar shells. Equator (mapped by trilobites) through Canada, Siberia, Greenland, Queensland. Ended in an ice age.
	500+	Cambrian	Small, shelly fauna appear, phosphatic replacement. 544 Cambrian Explosion: all main phyla appear. Animals diverse, increase in size beyond millimetres. Rise of predators and animal interactions (evidence: 520 Burgess Shale, 530 Chengjiang, 540 north Greenland).
	Earlier	Precambrian	Many groups of small animals (evidence: 560 Ediacaran, 680 south China).
	When?		First animals.
	3500		Oldest microscopic fossils. Tiny puffs of oxygen.
	3800		Earliest preserved rocks, Greenland.
	4600		Earth spun off sun, plus meteorite accretions: carbon.

Fig. 2.2 The geological succession.

of living animals show that common predatory gastropods attack barnacles at these junctions. The plate reduction coincided with the rise of this family of predacious gastropods in the Cretaceous. Further, the one lineage with no plate reduction is *Chenobia*, barnacles living on turtles and free from gastropod predation.

2.4 Can the fossil record date the earliest appearance of animals?

The difficulty is that all the main phyla seem to have appeared over a relatively short period of time in the Cambrian. Modern methods (based on the rate of breakdown of uranium to lead) date the 'Cambrian Explosion' as beginning 544 million years ago (Ma) and lasting for 5 to 10 million years. Our knowledge of early animals has been derived mainly from five very rich fossil sources; in order of increasingly distant past time, they are:

The Burgess Shale in Canada (520 Ma). The steeply sloping hillside leading down to a toxic, oxygen-depleted sea bed probably caused animals to be buried rapidly and protected at the bottom from scavengers and from bacterial decomposition. This fossil assemblage, first discovered early in the twentieth century but only recently fully analysed, has more than any other formed our picture of the Cambrian fauna.

Chengjiang in southwest China (530 Ma) has recently filled in that picture, with many of the same animals in different proportions. In particular, the well-preserved arthropods have given us a firm idea of their early diversity.

Sirius Passet in north Greenland (540 Ma) more recently still has among other riches provided fossils of what may be stem arthropods. This suggests that arthropods originated in the Precambrian.

The Ediacaran Range in South Australia (560 Ma) has been known since the mid twentieth century. These Precambrian (now Lower Cambrian, the border moves back) fossils were originally thought to have evolved quite separately from the Metazoa. They were called 'vendobionts', after the Precambrian Vendian period. It is now clear that they are not separate but consist of many early metazoan forms, cnidarians and a number of other phyla such as molluscs and various worms. Probably they were all soft-bodied, depending for their food on symbiotic micro-organisms (see Box 16.1).

The Yangtse Gorge in south China (580 Ma and earlier) has phosphorite deposits that recently have provided us with most striking early fossils. In addition to undoubted sponges, revealed by cellular imprints as well as spicules, there are early stages of embryonic cleavage, two- and four-cell stages and later ones with many blastomeres.

Was there a 'Cambrian Explosion'? That is a correct description of the fossil record as we have it. Yet the different phyla (or groups of phyla) must have separated considerably earlier, probably in the form of animals too small to be recognisable among the single-celled fossils known from the very earliest rocks. If there was an 'explosion', why did it occur? Explanations dwell on events such as increasing oxygen supply; more possibility of making collagen and hard structures; the increase in interactions between species, in particular the rise of predators. Or the change may simply have been due to the appearance of rocks able to preserve fossils.

Dating origins is very controversial at present. These basic relationships are difficult to resolve by molecular methods, which suggest much earlier dates for the establishment of animals than fossil evidence can verify. A study of 18 genes placed the division of the main groups of animals as early as 670 Ma, with the earliest divergence between the earliest animals and these main groups 1000 Ma or more. Is this discrepancy because molecular change may not accurately measure the passage of time, or is it because earlier fossils have not been found? Evolutionary lines could have separated long before morphological differences became detectable in fossils; we await more evidence. At present fossil evidence starts life too late, perhaps due to missing fossils, and molecular evidence starts life too early, perhaps due to statistical bias. The two estimates are converging, but seem unlikely to meet firmly.

In conclusion, clearly fossils must be included in phylogenetic reconstruction, although cladistic analysis of fossils is especially vulnerable to lack of information about characters and to underestimation of convergence. We do risk overemphasising characters fossilised by chance but, as with any method, we have to work with what we have got. We do risk interpreting fossils according to our preconceptions but, again, we can do that in any form of biological enquiry. At least we can now compare the evidence from three sources (fossils, morphology and molecules) to assess the relationships of present-day animals.

2.5   How can we use molecules to trace phylogeny?

Recently, interest in phylogeny has received a new impetus, as our rapidly growing knowledge of genetics and molecular evolution is providing a new approach, a new source of evidence about the course of evolution.

Molecules can be used in different ways. For example, theoretically, the total genes of two different species can be compared. The underlying idea is that if the genetic difference between two species is slight, they are liable to be closely related, and the degree of genetic difference will indicate the closeness of that relationship. The practical difficulties are enormous because most animals have so many genes, and there are theoretical difficulties also. To base our study on the genotype itself seems very attractive, but we must as always be wary of reduction to the simplest units, as this will eliminate essential information that depends on the organisation and interaction of those units. The sheer presence of particular genes does not define their effects: the action of genes depends upon other genes present, and a small change in genes can make a big change in animals. That we share about 99% of our genes with chimpanzees at once illustrates this point. Even though 1% may be a large number of genes, we are not as similar to chimpanzees as the figures suggest. As will be explained, the rate of genetic change is not the same as the rate of species change, and neither rate has remained constant during evolution.

Using particular molecules as though they were morphological characters presents fewer difficulties: here the idea is that if a molecule (a gene or the immediate product of a gene) changes slowly during evolution, comparison of the amount of change in that molecule will reveal the closeness of the relationship of different groups of animals. For example, a particular molecule might be compared in a range of insects, shrimps and spiders. Comparison should first show that different insects resembled each other more closely than any of them resembled shrimps or spiders, and might then indicate which two of these three groups are most closely related. In short, molecular characters can be substituted for morphological characters to assess the relatedness of animals, and it is this use of molecules that is introduced in the present chapter, to be discussed further in Chapter 20.

2.6   Which molecules are used?

Early work used the proteins which are produced by gene action, but recent work has concentrated on DNA, the genes themselves.

2.6.1   Ribosomal DNA (rDNA)

The genes coding for the RNA of ribosomes (rRNA), most often the small subunit 18S rRNA, are used (the gene is called 18S rDNA). This gene is very highly conserved, i.e. has changed very slowly in evolution, no doubt because it has an important structural role and mutations are unlikely to survive natural selection. It can therefore provide evidence about changes that occurred very early in animal evolution, such as the separation of classes within a phylum or even the origin of new phyla.

2.6.2   Genes regulating early development

These may be very informative, as genes acting early in development order the fate of whole blocks of cells. Comparisons between phyla are therefore based on molecular evidence quite different from that obtained using ribosomal genes, where information is obtained about the amount of sequence divergence that has occurred over evolutionary time. Using more than one gene and testing different aspects of evolutionary change may avoid some of the disadvantages inherent in using ribosomal genes alone (see below).

2.6.3   Mitochondrial DNA (mtDNA)

The genes situated in the mitochondria, outside the nucleus and different from the nuclear genes, are useful sources of information. In most animals the sperm contributes no material to the zygote, and therefore mitochondrial inheritance is confined to the female line: the use of mtDNA is simplified by its uniparental origin. Mitochondrial DNA changes relatively fast in evolution and is useful at the short end of the evolutionary time scale, to resolve changes that occurred less than 15 million years ago such as the separation of genera and species. When the whole (small) genome can be sequenced, mtDNA can also tell us about ancient changes. More often, it is not the gene content of the mitochondria that is used but the order of genes round the chromosomal ring.

2.7 How is molecular information obtained?

2.7.1 Gene products
The differences between proteins can be revealed and estimated using gel electrophoresis, a technique invented before the genetic material itself became accessible for direct study. Techniques applied to DNA directly are now more commonly used.

2.7.2 DNA hybridisation
The paired strands of DNA dissociate when heated because the bonds between corresponding nucleotides are broken. They recombine on being cooled. Single strands of DNA from related species can be put together and bonds will form, but only at the sites which correspond. When such a 'hybrid' is carefully heated it dissociates at a temperature lower than that required to dissociate perfectly matched DNA, because fewer bonds will have been formed. The difference between the two temperatures ('melting points') can be used as a measure of the genetic similarity of the two species.

2.7.3 Restriction site analysis
Restriction enzymes cut DNA at predictable sites into fragments about 4 to 6 nucleotides long. Fragments from different sources can then be compared, to obtain information about a small part of the total molecule.

2.7.4 Sequencing of nucleotides
This process allows the identification of each nucleotide in the whole sequence of a DNA molecule. This exhaustive process has been made easier by the polymerase chain reaction (PCR), which amplifies a small quantity of material for rapid analysis, and by the automated sequencing machine. This technique opened the way to the present explosion of molecular information.

2.8 How is molecular information processed?

Molecular evidence provides a large number of characters, all precisely defined. A character is usually a given nucleotide at a given site on the DNA molecule. Can the methods used for morphological characters be applied?

The traditional method of assessing characters is clearly not applicable.

Phenetic analysis can be applied to molecular differences. If molecular change increases at a constant rate as evolution proceeds, the amount of change is a measure of evolutionary distance. However, genes do not always change at a constant rate (see below).

Cladistic analysis is generally used, and some of its drawbacks disappear, because the characters are precisely defined and equivalent, and sufficiently numerous for statistical analysis to be substituted for parsimony. Selection of an outgroup to root the cladogram is, however, even harder.

2.9   How reliable is molecular taxonomy?

This is a large question that has been very controversial, but as more and more animals are studied using an increasing number of different molecules, confidence in molecular taxonomy is growing fast also. An indication of the advantages and disadvantages of using molecules may still be useful.

2.9.1   Advantages

1. The equivalence of data, since the nature and position of the unit is precisely defined.
2. The enormous size of the data set.
3. Statistical analysis of cladograms, avoiding the pitfalls of parsimony, is possible and only awaits agreement on the statistical methods to be used.
4. Where change in a molecule is rare, as in genes coding for ribosomal RNA, it becomes possible to trace relationships far back in time.
5. Non-heritable variation is avoided.

2.9.2   Disadvantages

1. The underlying assumption for most methods is that change in a gene molecule will depend only on the mutation rate and the time elapsed, i.e. that an unvarying 'molecular clock' is ticking at a regular rate. However, the clock is known to be variable in certain conditions, and the whole idea of functionally neutral changes in genes is controversial. Some branches of the evolutionary tree are known to evolve very fast: should we compensate by a subjective decision to omit such species (or groups of species) from our calculations? This has frequently been necessary to obtain results from ribosomal genes.
2. There is no record of past changes in characters. This is a serious disadvantage, as there are only four possible

nucleotides for any site in the DNA molecule. If there have been changes from one nucleotide to another and back again, such 'multiple hits' cannot be detected. This also is primarily a problem with ribosomal genes.

3. There is no recognisable intermediate condition between characters and, worse, no primitive condition for a given site can be recognised.

4. Functional correlates of character change can very seldom be traced.

5. It is very difficult indeed to root a tree derived from molecules; sequence similarity is the only guide, and the likelihood of convergence is usually impossible to assess.

For all these reasons, the need to use several (ideally many) different genes is apparent.

2.10 What is the present state of phylogenetic enquiry?

Ever since Darwin, biologists have wanted to understand the evolutionary relationships between groups of animals. Sources of evidence have included the fossil record and the study of animal development (ontogeny) as well as morphological comparisons. In recent years there has been rapid progress, due to cladistic analysis of morphological characters, new fossil discoveries and new understanding of the genetic basis of development, with the use of molecular characters as an entirely new source of phylogenetic evidence.

Molecular characters are not better than morphological ones but they are different. They are copious and comparable with each other. Their primary value is in providing an independently derived phylogenetic tree for comparison with phylogenies based on morphology. As has been explained, it is necessary to compare the results from a number of different genes; there are now many nuclear genes in use and mitochondrial genes are being used in new ways. Molecular and morphological evidence sometimes suggest quite different patterns of evolution, but increasingly often the results coincide. When molecular data coincide with one morphological tree rather than another, this is strong evidence for the correctness of that tree.

With this success, quite a new problem has arisen: zoologists are so excited about tracing phylogenetic relationships that the animals themselves get neglected! This book attempts to balance the study of invertebrates. It now is time to describe the animals themselves, to explain the ground plan of each phylum and to indicate the evolution that has occurred within it. Our present understanding of relationships between phyla is then summarised in the final chapter (20).

Chapter 3

Porifera

Sponges are by far the simplest multicellular animals and are very different from all the others. They have no fixed body shape, no plane of symmetry and are covered in holes. All sponges live in water, nearly all in the sea. The cells are uncoordinated, cell differentiation is entirely reversible and cells may wander about in the background jelly. A whole sponge can be regenerated from a few separated cells. Sponges can almost be regarded not as individuals but as colonies of separate cells; almost but not quite, as most have a skeleton made of spicules that supports the body.

These very simple animals are nonetheless very successful and widespread: since the early Cambrian they have covered most of the suitable surfaces on the shore and in the shallow sea: the latest survey found 15 000 living species. How is it that such simple animals can do so well? What has there been for natural selection to work on in this phylum? How fundamentally do they differ from other animals and what are their evolutionary and ecological relationships with them? To address these questions, we must study the basic structure and the different kinds of sponges, and indicate the ways in which they make a living.

3.1 What are the distinguishing characters of sponges?

Sponges are sessile and immobile, having neither nerves nor muscles. There may be slight contractility round the larger pores but it is very restricted.

Sponge cell types are the distinctive collar cells or 'choanocytes' (Figure 3.1a), the 'pinacocytes' that make an outer layer and the 'amoebocytes' wandering through the central jelly or 'mesohyl'. This jelly is needed for support; in contrast to other animals, neighbouring cells are not bound together by a basement membrane.

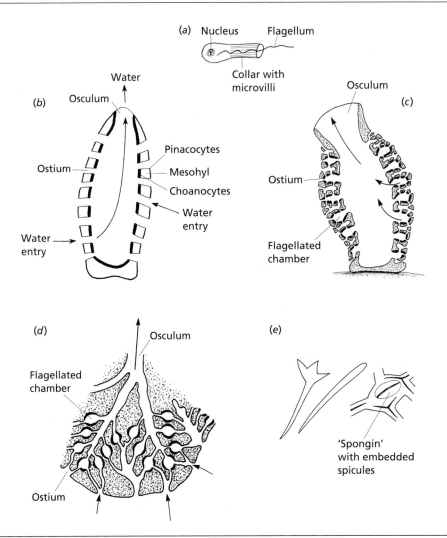

(a) Nucleus   Flagellum
Collar with microvilli

(b) Water
Osculum
Ostium
Pinacocytes
Mesohyl
Choanocytes
Water entry
Water entry

(c) Osculum
Ostium
Flagellated chamber

(d) Osculum
Flagellated chamber
Ostium

(e) 'Spongin' with embedded spicules

Fig. 3.1 The structure of sponges: (a) a choanocyte, component of black regions in (b), (c), (d); (b) basic sponge structure, as in the late larva; (c) folding of walls to make flagellated chambers, for example in *Leucosolenia*; (d) part of *Grantia*, a fully elaborated sponge; (e) spicules.

Small pores perforate the whole body (the name of the phylum, Porifera, means 'pore bearing'). Water carrying food particles enters the body by many small pores ('ostia'), moved in by the beating of the flagella of the internal collar cells. These cells extract food particles from the water, which flows out through larger pores, the 'oscula' (Figure 3.1b). The structure becomes elaborated during evolution, as in Figure 3.1c,d. This arrangement, with the principal openings exhalant, is unique to sponges.

The skeleton is made of spicules (Figure 3.1e) of calcite (a calcium salt) or silica (a silicon salt) with or without a matrix of horny collagen-type protein. Such use of silica is a unique feature. An unusually wide range of skeletal materials occurs among closely related species of sponges. Animal skeletons, whether hard or hydrostatic (based on the incompressibility of water) usually translate muscle contraction into movement. Sponges, however, have no muscles and the supporting skeleton instead serves to prevent movement from occurring. Spicules have further important functions in preserving the sponge's shape, keeping the pores open and maintaining the internal channels (as well as making the sponge even nastier to eat than it probably would be anyway).

3.2 What different kinds of sponge are known?

Calcarea, with calcareous spicules. They occur in shallow waters (less than 100 metres). Examples are *Leucosolenia* and *Grantia* (Figure 3.2b,d).

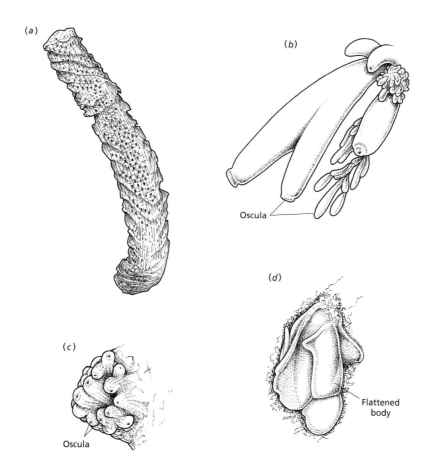

Fig. 3.2 Drawings of (a) a hexactinellid, showing spicules fused to form a lattice; (b) *Sycon* (larger) and *Leucosolenia*; (c) *Halichondria*; (d) *Grantia*.

Demospongiae, with siliceous spicules. Examples are *Halichondria*, the breadcrumb sponge (Figure 3.2c), and *Spongia*, the once commonly used bath sponge, in which there are no spicules but only protein fibres.

Hexactinellida, the 'glass sponges' (Figure 3.2a), are very different from other sponges. They consist of a small group (400−500 species) in the deep sea (below 200 metres); their growth is orientated to the constant water currents there. The tissues are 75% syncitial (lacking cell boundaries) with the remaining cells connected by cytoplasmic bridges; even the choanocytes are not separate cells. There are no pinacocytes and no cells with any contractility. The skeleton is a lattice made from six-rayed spicules of silica. The skeletal network is useful in distributing light, as spicules have fibreoptic qualities with refractive differences between the core and the shell of the fibres. Some authorities believe that the hexactinellids evolved separately, in parallel with other sponges, but there is increasing evidence that they are a specialised offshoot from cellular sponges: for example, the embryos resemble other sponges in their early development, with cell boundaries, but later hatch as syncitial larvae.

Subdivision of the classes has traditionally depended on spicule structure, but more recently biochemical or reproductive characters, or those discovered by electron microscopy, have been preferred. For a time a further class, the 'Sclerospongiae', was proposed for some 15 species of sponges where spicules have become fused to make massive skeletons, building coral-like reefs, since the Palaeozoic. This, however, is not a natural group; it includes some Calcarea and some Demospongiae that have evolved in parallel.

3.3   How do sponges make a living?

3.3.1   Feeding method

Choanocytes both create water currents and trap food. The collar is a ring of about 30 small folds ('microvilli', Figure 3.1a), linked by cross-bridges. Food is absorbed by food-engulfing cells at the foot of the collar. Within both the Calcarea and the Demospongiae the internal pattern has changed (Figure 3.1c,d) to provide canals and flagellated chambers. This arrangement provides a larger surface area covered in choanocytes and slows the flow of water past these cells, allowing more time for food capture.

Sufficient food particles arrive because water pumping is surprisingly intensive: for example, a specimen of *Leuconia* (Demospongiae) growing 10 cm tall with a diameter of 1 cm contained about 2.2 million flagellated chambers and pumped 22.5 litres of water per day. The oscular outward jet was 8.5 cm per second. Amoebocytes assist the circulation of food through the sponge: all digestion is intracellular. As in many permeable marine animals, dissolved organic material may be a subsidiary source of food.

3.3.2 Behaviour

The animal is sessile, there are no sense organs, nerves or muscles; what can there possibly be in the way of behaviour? Starting with the lowest expectations, we can find quite a bit. For example, the rate of flagellar beat can be influenced by currents: in one experiment with *Halichondria* the flagella beat at 3 cm per second in still water and this rose to 7 cm per second as the external current was increased. There is a measure of communication, even coordination, between cells: for example, dilation of a channel may be propagated and stimuli such as touch, exposure to air or poisons can result in the closure of a distant osculum. Although there are no tissue junctions between cells there may be communicating channels, rapidly and temporarily formed. Reactions are slow, as is shown by *Hymeniacidon*, a common encrusting orange growth, about a centimetre thick, on British beaches. Poke it, and about 10 minutes later the osculum will close. It is not clear how this contractility is achieved, but occurring round the oscula there are some amoebocytes called 'myocytes', that are particularly rich in microfilaments and microtubules. There is some evidence that myocytes may contain the fibrils (actin and myosin) which are the basis of contraction in all other animals investigated (and even in some unicells). Nor is it clear how cells can be re-extended, except by the pull of neighbouring cells, but water pumping must help to retain the shape of the sponge. After all, a sponge does not need rapid reactions: it needs only to close up fast enough to avoid desiccation when the tide goes out. A worse hazard would be to close the exhalant osculum while the flagella continued to beat: the sponge might burst.

Sponge larvae are motile, using flagella. Their movements are not under any form of nerve control, but the cells respond to changes in light intensity, which can alter the direction of swimming. When first released the larvae swim upwards in the sea, rotating as they swim: increased intensity of sunlight from above stiffens the flagella of the rotating larvae, steering them to darker areas down below.

It is in the very different Hexactinellida that a greater degree of coordination has been found. Here the diameter of oscula cannot change, but on mechanical or electrical stimulation the flagella in all the chambers may stop beating. There are no nerves: electrical impulses pass along the continuous tissue of the syncitium. In other sponges no such conducted electrical signals are known.

3.3.3 Reproduction and development

Sponges have remarkable powers of regeneration: they can be strained through a fine mesh yet the cells will come together, aggregate and divide to reconstitute the sponge. Cells from different

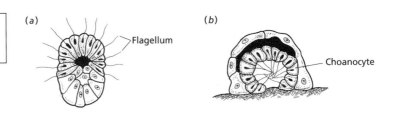

Fig. 3.3 Sponge development:
(a) amphiblastula larva;
(b) gastrulation after settling.

species will not aggregate. In the sponge body asexual reproduction by budding occurs readily: it is hard to distinguish from growth. Some freshwater sponges bud off parts of the body to form 'gemmules', stages resistant to adverse environments that regenerate when conditions are favourable.

In sexual reproduction, gametes are formed in the mesohyl by dedifferentiation of other cells. Most sponges are hermaphrodites, but cross-fertilised. When sperm of the same species enters through an ostium, it is engulfed by a choanocyte which loses its flagellum and moves through the jelly until it finds an egg, a procedure very different from that of other Metazoa.

Sessile adult animals always need free-swimming larvae for dispersal. Sponges have simple flagellated larvae, usually developing in the parent body and then freed to swim and settle in suitable sites (Figure 3.3a). Some species achieve further dispersal by asexual fragmentation followed by release of larvae from the dispersed fragments. Once the larva has settled the cells move and become rearranged in a process which corresponds to the gastrulation (see Glossary) of other Metazoa and is controlled by similar genes. The larva now has differentiated cells patterned along an axis (Figure 3.3b); this pattern becomes less clear at metamorphosis. In some sponges the outer flagellated cells are lost at gastrulation, but in the Calcaria they dedifferentiate into multipotent cells. The cells, whatever their derivation, then move, divide and differentiate.

3.4 What changes have evolved during sponge history?

3.4.1 Morphological change
In these simple animals, with so little connection between cells, there is not much scope for morphological evolution beyond the elaboration of flagellated chambers related to increased efficiency of water circulation (see above). Sponges show an array of growth forms according to environmental conditions: where it is exposed to wave action a species usually displays a flat encrusting growth, but in crevices or still water the same species may grow tall or hang down, increasing the surface area. Deeper in the sea there is more variety, including metre-thick spheres and the reef builders already mentioned. In turbulent water continual disturbance brings in

plenty of food, but in still water a sponge is in danger of recycling the water and gaining no more food. Increased body size is then an advantage, and a raised osculum will achieve a more powerful exhalant jet. A few of the simpler Calcarea can be recognised by their shape: for example Figure 3.2 shows the upright tubes of *Leucosolenia* and the flat 'purses' of *Grantia*, but usually the general appearance of a sponge is no certain guide to its identity.

3.4.2 Physiological differences

Such differences may be marked and may be correlated with different ecological niches. In the Demospongiae, for example, *Mycale* is an opportunist generalist readily colonising new sites but never growing to the maximum possible size; most of its energy is devoted to rapid growth and reproduction. *Tethya* by contrast forms permanent populations of large individuals in less favourable environments; its energies are channelled to physical resistance and it reproduces only slowly.

One extraordinary group of sponges has become carnivorous: living in deep-sea habitats where small particle food is scarce, they capture small crustaceans with Velcro-like raised hooked spicules. The crustaceans become entangled, grown over and gradually digested. These sponges are hydroid-like in form (see Chapter 4) and have entirely lost choanocytes, ostia, oscula and water channels. They can be recognised as sponges (Demospongiae) only by their spicules and by the nature of the outer layer.

However, the general form and function of animals so much constrained by structural simplicity did not give natural selection very much to work on. Diversity is shown far more strongly at the biochemical level.

3.4.3 Sponge biochemistry

Why are sponges so highly coloured? Not all their colours can be due to the commonly occurring symbiotic algae. On the shore and in the shallow sea they are often yellow, orange, red, green or violet. How can animals so openly exposed avoid predation, even if they do contain spicules? Why are they the only animals to be able to extract and build with silica? The answers to all these questions reside in unusual and very varied cell biochemistry. Colours may be due to pigment granules in amoebocytes and may serve as a warning of inedibility. Sponges produce an array of biotoxins that discourage predators; they may extend their use of poisons to chemical warfare with other sessile invertebrates, to compete for living space. One tropical encrusting sponge, *Terpios*, can grow as much as 23 mm per month as it poisons its neighbours. The family Clionidae (Demospongiae) includes 'boring' sponges with specialised amoebocytes whose chemical secretions remove calcareous fragments from coral skeletons, clams and scallops. The chips are collected into the exhalant currents within the sponge and pass out through the osculum. In coral reefs boring sponges gain protective shelter, and cause considerable damage.

Cell biologists and pharmacologists are currently very interested in sponges. They provide model systems for the study of cell junctions of the simplest and most labile kind, and for the investigation of cell surface proteins that mediate cell recognition in a very basic immune system. Sponges produce bioactive compounds, some of which may directly benefit us: for example a sponge long used by New Zealand Maoris to promote wound healing has been found to contain high concentrations of a potent anti-inflammatory agent. Antimicrobial action is often found; a growing list of such examples emphasises that we need to conserve the biochemical diversity of sponges.

3.5   How are sponges related to other phyla?

This question has aroused much speculation. For a time sponges were separated from all other multicellular animals as a subking-dom, the Parazoa, and stigmatised as a 'side issue'. This classification and the ideas behind it have disappeared. Sponges are now known to be Metazoa of the simplest possible kind, as expected since the cells are so loosely held together. The only other phylum with equally simple structure is the Placozoa, very small flat plate-like aggrega-tions of amoeboid cells (Figure 3.4a). They were once thought to be some developmental stage of a sponge, until sexually mature individuals were recorded. Placozoa are now placed in a separate phylum, with one genus, *Trichoplax*; recent molecular work compar-ing specimens from different sources has revealed that this 'genus'

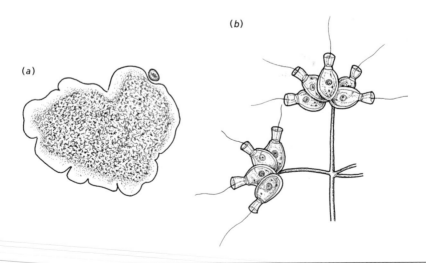

Fig. 3.4   Non-sponges: (a) *Trichoplax*, a placozoan (100 μm in length), superficially sponge-like; (b) a choanoflagellate colony, related to sponge ancestors.

in fact contains a great variety of forms. Placozoa are not at all closely related to sponges: their nearest relatives may be the Ctenophora (described in Chapter 5).

Sponges are known to be ancient: spicules fossilise readily and are very common in Cambrian deposits. They have recently also been found among the very earliest fossil animals in the Precambrian era (see Chapter 2). Recently discovered deposits in China contain not only spicules but prints of soft tissue, embryos and larvae from about 580 million years ago.

Sponges cannot be classed with the Cnidaria as having two cell layers (nor are they radially symmetrical) and they are even less like all other Metazoa. Yet molecular evidence strongly supports a single origin for the Metazoa, with sponges separated from the rest more than 600 million years ago. Tracing this single origin is clearly difficult, and no phylum can be derived from present-day forms, but choanocytes do resemble the unicellular Choanoflagellata, some of which form colonies (Figure 3.4b). Study of choanoflagellate proteins reveals the expression of a number of developmental genes which occur in sponges and other Metazoa but in no other organisms (see Chapter 20). Accordingly, the dominant hypothesis at present is that sponges did arise from ancestors shared with those of choanoflagellates, which are therefore seen as the sister group to all Metazoa.

3.6   How have sponges become so successful?

Morphological simplicity and lack of coordination have not prevented sponges from being extremely successful animals, if success is measured by survival, large numbers and very widespread distribution (in the sea). Sponges remind us that complexity of form is not the only route to success. Sponge diversity may be limited in morphological terms but their relatively independent cells have been able to evolve a variety of unusual biochemical specialisations.

Sponges are 'alternative' animals: they can respond to the environment and behave as functional units, but they do it in ways unique among multicellular animals.

Chapter 4

Cnidaria

Cnidaria include the anemones, corals, jellyfish and hydroids, i.e. all the animals formerly included with the comb jellies ('Ctenophora') in the phylum Coelenterata. They may be in the form of sessile polyps, or freely floating medusae (Figure 4.1a,b). All are aquatic, nearly all are marine, and they are very simple in structure. Yet there are vast numbers of individuals belonging to at least 10 000 species widely dispersed in the sea, varying in size from individuals a few millimetres across to coral colonies measuring hundreds of metres. Stinging cells (called 'cnidae' or 'nematocysts') are used for food capture and defence; they are unique to the phylum and diagnostic of it.

The combination of simplicity of structure with large numbers and considerable diversity provides the theme for this introduction to the phylum. Discussion is focused on how such simple animals can make a living and what features have enabled them to become so diverse. This chapter is relatively rather full, because the emerging picture forms an important background to the consideration of more elaborate animals.

4.1 Why do we regard Cnidaria as simple?

They have no head end. The mouth (which serves also as the anus) is the single opening of the only internal cavity, called the 'coelenteron', which is an enclosed part of the water in which the animal lives. The mouth is usually surrounded by tentacles where the stinging cells are concentrated. Radial symmetry allows food capture from all sides, but it may be secondarily modified in relation to particular functional needs.

There are only two cell layers, the ectoderm (also called the epidermis) and the endoderm (also called the gastrodermis). They are separated by a jelly-like 'mesoglea', which contains some cells and connective tissue fibres but is not itself a cell layer (Figure 4.1c). Accordingly Cnidaria are said to be 'diploblastic', in contrast to

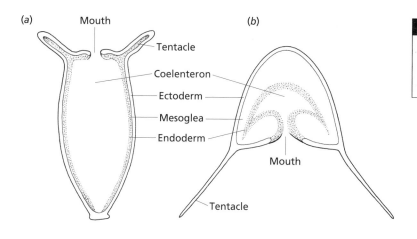

(a)

Mouth

Tentacle

Coelenteron

Ectoderm

Mesoglea

Endoderm

(b)

Mouth

Tentacle

Fig. 4.1 Cnidaria: (a) polyp and (b) medusa forms; (c) longitudinal section of *Hydra* to show cell types scattered in the two layers.

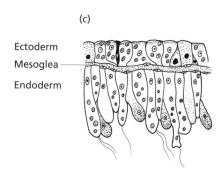

(c)

Ectoderm

Mesoglea

Endoderm

all other multicellular animals (except sponges), which are 'triplo-blastic', having three cell layers. Figure 4.1c shows that cells of the same type are not arranged together in either layer (the slightly confusing description of Cnidaria as having 'tissue grade' organisation emphasises that there is no aggregation of tissues to make organs). There is no brain or central nervous system, but a network of multipolar nerve cells conducts slowly in all directions (Figure 4.2a). There are no separate muscles but 'musculo-epithelial' cells of the ectoderm and endoderm are drawn out at the cell-base into contractile muscle tails that extend up and down or around the animal in the mesoglea. These muscle tails form sheets that may then be folded to make compact structures. This is a unique method of making solid muscles (Figure 4.2b,c).

As all cells in both layers are directly in contact with environmental water, either on the outside of the animal or in the coelenteron, there are no special structures for respiration and excretion, nor is there a transport system (apart from sea-water channels in some large jellyfish).

Cell movement is a further distinctive property of Cnidaria. Not only are there migratory stem cells (called interstitial cells) that give rise to the nematocysts, nerve cells and gonads, but also apparently more differentiated cells may continuously undergo cell division and move. This capacity is most marked in hydrozoan polyps;

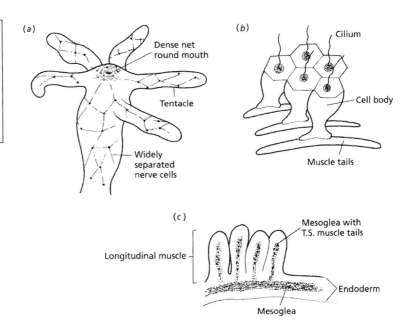

Fig. 4.2 (a) The nerve net of *Hydra*; (b) muscle tails on the musculo-epithelial cells of a sea anemone, *Metridium*; (c) transverse section of mesentery showing folding of mesoglea with muscle tails, making a longitudinal 'muscle.'

in *Hydra*, for example, ectodermal cells are continuously produced just below the mouth region and migrate to the tentacle tips or to the foot, where they are sloughed off. At a certain distance from the dominant tentacular region, cells can move out and divide to form a bud. This ability to move and differentiate underlies the remarkable regenerative powers of many polyps; an isolated piece of tissue can often regenerate a polyp with one or more tentacular regions, even after having been put through a sieve, in strong contrast to what happens with more elaborate animals.

The simplicity of cnidarians is thoroughly established by their diploblastic constitution, the paucity of cell types, the nature of the only internal cavity, the lack of organs and the absence of separate muscle cells and centralised nervous systems. The unique stinging cells (Figure 4.3) make these simple animals viable.

4.2 What kinds of Cnidaria are known?

There are four classes. The range of structure in cnidarians is shown in Figures 4.4 to 4.8.

Anthozoa: anemones and most corals. Polyps with vertical divisions (mesenteries) in the coelenteron (Figure 4.4). No medusa forms.

Alcyonaria, polyps with eight mesenteries and typically eight branched tentacles: sea pens, branching corals, soft corals.

Zoantharia, typically with six or twelve mesenteries and variable numbers of simple tentacles: anemones and 'true' oceanic reef-building corals.

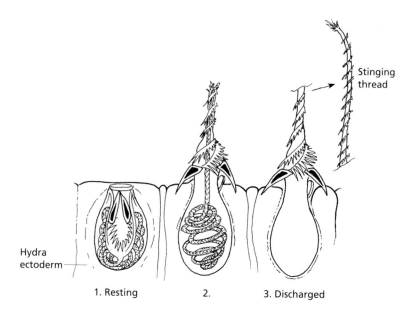

Fig. 4.3 Nematocyst discharge.

Stinging thread

Hydra ectoderm

1. Resting   2.   3. Discharged

Scyphozoa: jellyfish. Medusae with the mesoglea much expanded; no velum. The medusa form is dominant but may develop from a transient polyp-like sessile stage. Jellyfish are very common in all oceans; they may penetrate shallow seas or be washed ashore.

Cubozoa: box jellies. Medusae with four sides and a marginal shelf, or velum. Found in tropical seas, e.g. off Queensland, Australia, where they are known as 'sea wasps' and can be lethal to humans. They are very different from scyphozoa, notably in having most elaborate eyes (see below).

Hydrozoa: hydroids, typically with both polyp and medusa stages in the life cycle. The polyps are small, without mesenteries in the coelenteron, and the medusa may have a velum. Polyp colonies are often intertidal; some groups are oceanic. *Hydra* lives in fresh water. The many forms include:

    Colonial polyps (a few are secondarily solitary, e.g. *Hydra*)
    Trachyline medusae (oceanic, without polyps)
    Milleporine corals (see below)
    Siphonophora (highly specialised polymorphic colonies).

4.3 How do Cnidaria make a living?

4.3.1 Nematocysts and feeding

Nematocysts (Figure 4.3) are the stinging cells that make it possible for these sessile polyps and floating jellyfish to be predatory carnivores, often feeding on animals larger than themselves. Nematocysts are specialised cells borne mainly on the tentacles.

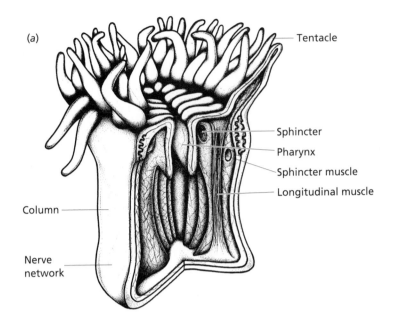

(a)

Tentacle

Sphincter

Pharynx

Sphincter muscle

Longitudinal muscle

Column

Nerve network

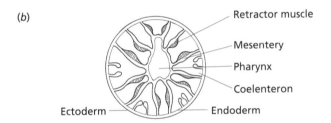

(b)

Retractor muscle

Mesentery

Pharynx

Coelenteron

Ectoderm

Endoderm

Each consists of a closed capsule covered by a flap and containing a coiled barbed thread or tubule filled with paralysing toxin. Some nematocysts do not contain toxin; instead, the thread entangles small prey or adheres to it by the barbs.

Nematocysts are triggered by a combination of chemical and mechanical stimuli, causing the capsule flaps to open with a sudden release of bound calcium ions, which leave the capsule. Water then rushes in by osmosis. The greatly increased hydrostatic pressure inside the capsule, aided by release of stored energy in its walls, results in explosive discharge; the tubule is everted and toxin released from its tip into the prey. All this occurs within a fraction of a second. The used thread is discarded and interstitial cells move into the space to form more nematocysts. When capturing a brine shrimp *Hydra* may lose a quarter of its nematocysts, but it replaces them within 48 hours.

Nematocysts can react quite independently, and as the whole mechanism is contained in a single cell they can react very fast. However, they are often coordinated by the nerve network, producing bursts of nematocyst discharge.

Nematocysts can also be used for defence against predators, and to attack rival polyp colonies competing for space. In some species of anemones and corals, contact with a foreign clone stimulates nematocyst discharge and the resulting 'stinging war' spaces out the different clones.

Nematocysts are not known in any other metazoan animals except for nudibranch molluscs, which extract them from their cnidarian food and, remarkably, harness them for their own use. A clue to the possible evolutionary origins of nematocysts comes from the spores of intracellular parasitic protista. Microsporidia have spores containing coiled eversible threads and Myxosporidia have spores containing nematocyst-like capsules with hollow eversible tubules. When in contact with a suitable host, the spore everts its tubule and through it discharges the entire spore contents into the host cell. Perhaps an early cnidarian ancestor was able to incorporate and use some such parasite; but that is pure speculation.

4.3.2   Nerve conduction

Response times in Cnidaria are slow, constrained by the structure of the nerve net and delay at the nerve-cell junctions. Nervous centralisation is totally lacking. Where there are many nerve cells close together, as in the mouth region of a polyp, this is a region of particularly slow conduction rather than a rudimentary brain, as there are more junctions between cells, delaying conduction (Figure 4.2a). Apart from using the nematocysts, sessile polyps cannot make much response to stimuli and have no elaborate sense organs. Their nerve endings may be sensitive to chemical or mechanical stimulation, or to light, but it may be as much as 2 minutes before a cnidarian can react to a stimulus. Faster reaction occurs in some anemones (such as *Metridium*, often carried about on the shells of hermit crabs) where the column (Figure 4.4a) has 'through conduction tracts' in the form of bipolar nerve cells up to 7 or 8 mm long. Here the first impulse 'facilitates' conduction so that subsequent impulses are not delayed at all. Again, some medusae have a nerve net in the 'umbrella' rim that stimulates muscle to pulsate rhythmically, and both Cubozoa and some hydrozoan medusae have closer links between nerve cells in the nerve ring. Perhaps the limit to coordination is that further centralisation would not help a radially symmetrical animal.

4.3.3   Movement and locomotion

Cnidarians can move parts of their bodies (mainly the tentacles) and change the body shape. Locomotion (movement from place to place) occurs in medusae and also in a few polyps. Movement by muscle contraction requires a skeleton, because muscles must have some restraint to pull against in order to have any effect, and after contracting they must be re-extended, but the skeleton need not be hard. Like other soft-bodied animals, Cnidaria use

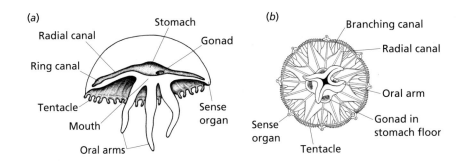

Fig. 4.5 The structure of a jellyfish (scyphozoan): (a) side view; (b) oral view.

the incompressibility of water to serve as a hydrostatic skeleton. Hard structures exist in Cnidaria, such as the calcareous skeletons of corals, but they never play any part in muscle contraction. Anemones (Figure 4.4) and jellyfish (Figure 4.5) provide examples of cnidarian movement: how far this can ever be termed 'spontaneous' is largely a matter of definition, but certainly anemones frequently contract and alter their shape in the absence of any apparent change in conditions. A few anemones may shuffle, burrow or swim, but most are sedentary. They contract their muscles against the water-filled coelenteron, expelling water through the mouth: at rest the invaginated sleeve or 'pharynx' acts as a valve which keeps water from escaping. Refilling the coelenteron is slow; it depends on ciliated grooves or 'siphonoglyphs' running down from one or two points on the pharynx circumference. (This, incidentally, is an example of functional needs interrupting radial symmetry.)

Compared to the slow reactions of anemones, jellyfish can swim relatively fast, as their muscles work against the much expanded mesoglea, which is elastic and springy. This 'jelly' is largely water; in solution the heavy sulphate ions of sea water are largely replaced by lighter anions, so that the mesoglea also provides the buoyancy that enables the animal to float. Storage of oxygen is another function of the gel, assisting vertical movements and survival at low oxygen levels. The bell pulsates rapidly due to the contraction of the striated muscle situated around its rim, and a group of nerve cells controls this rhythmical pulsation.

Rapid locomotion is achieved only in the Cubozoa, little known until recently. In an aquarium, it is fascinating to watch these animals with a seemingly minimal bell trailing four tentacles (perhaps 75 cm in length), the bell gently pulsating as they drift down the cylindrical tank and pushing them up again with a single quick contraction. Unlike other jellyfish, Cubozoa 'sleep' at the bottom of the sea at night. They are fast active predators with unusually well-developed eyes, able to hunt fish and to kill them with the very potent toxins in their nematocysts. They live close to the shore, and their very specialised vision may relate to avoidance of obstacles as well as to efficient food capture. They have many eyes in four groups, connected to the tentacle bases by a nerve ring, but there is no coordinating brain; instead, each eye sends impulses direct to the pacemakers for swimming. In each group most of the eyes are simple light detectors but two are remarkably well developed. These eight 'camera eyes' each have lens, retina and cornea, forming detailed colour images, which, however, only come to a focus behind the retina. The functioning of these eyes is not understood. There are eyes even in the (unusually simple) larvae; there is no nerve net in these larvae but a flagellum is directly rooted in each eye cup.

4.3.4 Reproduction

This may be asexual: as described earlier, *Hydra* polyps bud and the buds separate from their parent. Without separation, the same process can result in the growth of a colony. More commonly reproduction is sexual, achieved by the release of eggs and sperm into the sea.

4.3.5 Life history

Anthozoa have no medusa in the life cycle. The polyps release the gametes, which fuse to form a zygote and this develops into a ciliated swimming 'planula' larva, very simple in structure, which later settles and grows into the adult polyp. Some hydrozoans are similar (e.g. the freshwater *Hydra*) but characteristically the polyp colony, living in the harsh conditions of the intertidal region, buds off medusae which produce gametes and the zygote develops into a planula which grows into a new polyp colony (Figure 4.6a). This is not the 'alternation of generations' familiar in plants, since polyp and medusa have chromosomes identical in number and kind. Dispersal is of the utmost importance to a sessile animal, and the motile medusa provides a more robust agent of dispersal than the fragile planula larva. In Scyphozoa the gamete-producing medusa is dominant in the life cycle. The zygote develops not into a planula but into a sessile larva which buds off little jellyfish (Figure 4.6b). Some Cubozoa, most remarkably, can transform a polyp directly into a medusa. Such a transformation is highly unusual but not unique: medusae of the hydrozoan *Turritopsis* can revert to polyps.

4.3.6 Early development

Cnidaria, unlike other phyla, have a bewildering variety of developmental mechanisms. Such plasticity may be attributed to the basic simplicity of the structure and to the lack of constraints on cell movement.

Differences in growth and development are found between closely related species and even between individuals within a species, depending on environmental conditions.

Fig. 4.6 Two cnidarian life cycles; (a) the hydrozoan *Obelia*, showing the structure of the polyp colony; (b) the scyphozoan *Aurelia*. Both show the planula larva.

(a)

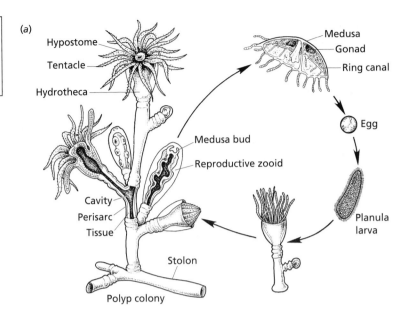

(b)

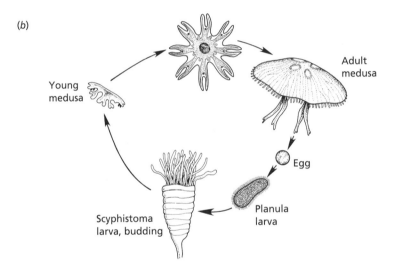

4.4 How has so much diversity been possible?

The diversity of Cnidaria is shown by three attributes:

1. **Polyp and medusa forms**. These two forms provide the basic diversity of Cnidaria, since they make possible two different ways of life, a sessile polyp or a free-swimming medusa. As has been explained, many cnidarians combine both forms within one life cycle.

2. **Colony formation** is the commonest result of the remarkable budding power of many polyps, allowing increase in size and the possibility of division of labour. Size increase is a great advantage in that a larger area can be swept for food. While individual medusae can enlarge by expanding the inanimate mesoglea (the largest jellyfish known are up to about two-thirds of a metre in diameter) individual polyps cannot get very big without supporting structures and increased digestive area; anthozoans with mesenteries dividing the coelenteron can become larger than hydrozoan polyps. In a colony, however, many small hydrozoan polyps can sweep as much water as a single anemone, and food is shared as the units remain in contact through a common coelenteron. They are limited in size by problems of mechanical support rather than by physiological constraints.

The commonest example of division of labour is that between the feeding and reproductive individuals in a hydrozoan colony. There may be further differentiation, for example between feeding and stinging individuals in the colony. Polymorphism is carried much further in *Hydractinia*, a colonial hydroid attached to the shell of a gastropod mollusc which is inhabited by a hermit crab. The planula larva settles on the shell and develops into a primary polyp, which buds into a mat of branching stolons (tubes of tissue with a coelenteron that connects polyps). These stolons bud off four kinds of polyps, all with batteries of nematocysts: first the feeding polyps, the only kind to have a mouth surrounded by tentacles; then the reproductive polyps, which bud off medusae that are not set free but release eggs or sperm into the sea; polyps with 'fingers' are formed only at the mouth of the mollusc's shell, where they extract eggs; development of the fourth type of polyp, with a single long tentacle for defence, is stimulated by the presence of foreign organisms. The stolons (unusually for stolons) contain a nerve net implicated in the coordination of polyp behaviour.

Polymorphism reaches its height in the floating colonies of the Siphonophora, which consist of highly modified individuals where the distinction between polyp and medusa is obscured (Figure 4.7). The division of labour between individuals in these colonies resembles that between organ systems in more highly organised animals. Cnidarians do not have division of labour between organ systems in a body, but the functional alternative in siphonophores is this polymorphism between genetically identical individuals in a colony.

3. **Coral formation** enormously extends the range and size possible for colonies. Any polyp that lays down calcium carbonate becomes a coral. Occasionally corals are solitary,

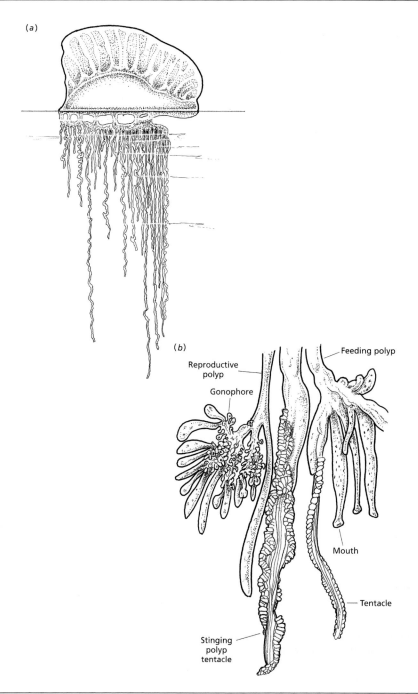

(a)

(b)

Reproductive
polyp

Feeding polyp

Gonophore

Mouth

Tentacle

Stinging
polyp
tentacle

Fig. 4.7 (a) *Physalia*, the Portuguese man-of-war, a siphonophore; (b) part of a *Physalia* colony, showing the division of labour between individuals.

for example the Devonshire cup coral, but nearly all are colonial. They occur in three groups of Cnidaria (Figure 4.8):

Milleporine corals (Hydrozoa). The skeleton is laid down outside the ectoderm of each polyp. The structure is never very large: it can be recognised by the two sizes of holes, large for feeding polyps and small for stinging ones. Medusae are produced at intervals (unlike anthozoan corals).

Alcyonaria (Anthozoa). The skeleton is laid down in the mesoglea by ectodermal cells. It may then be compacted, as in the organ pipe coral, or there may be a central supporting rod made of protein or of coral itself, as in the red coral used in jewellery.

Madreporaria (Anthozoa). The skeleton is laid down underneath the polyps; living tissue from the polyp base extends into the skeletal mass, leaving its imprint between the calcareous septa. These are the 'true' reef-building corals.

Perhaps the remarkable variability of Cnidaria relates in part to the unusual lack of constraint on cell movement and differentiation, but what has allowed, for example, the extraordinary complexity of the Cubozoa and the degree of coordination of *Hydractinia* colonies? The very recent discovery that unlike elaborate invertebrates these primitive animals have amazing genetic complexity, almost as much as ourselves, must surely be relevant.

4.5 What is the ecological importance of coral reefs?

Reefs built by corals are the largest structures ever made by any animals, including ourselves. Deposition of large quantities of calcium carbonate by a colony of many small polyps depends on the presence of green algae, living inside the cells of the coral polyps; such an association for mutual benefit is called 'symbiosis'. Photosynthesis by the algae provides the coral with some of its food and the algae assist uptake of nitrates and phosphates. Also, photosynthesis facilitates calcification by removing carbon dioxide, promoting the dissociation of calcium bicarbonate dissolved in sea water, with formation and precipitation of calcium carbonate:

$$Ca(HCO_3)_2 = CaCO_3 + H_2O + CO_2$$

Coral reefs occur in tropical seas. They need sufficient light for photosynthesis by the symbiotic algae and a temperature range of $23-29\,°C$ (with an optimum of $26-27\,°C$). These two factors also determine the depth at which living coral polyps can grow. This is

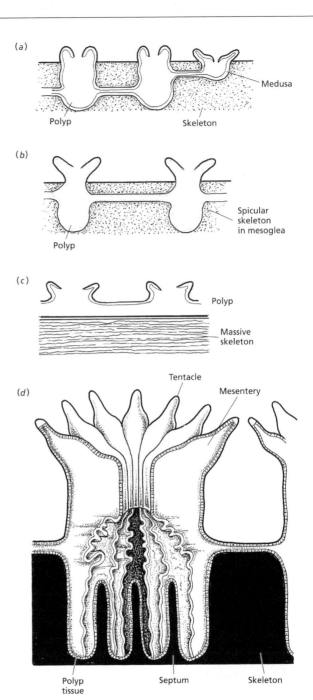

Fig. 4.8 Diagrams of coral structure: (a) milleporine coral, calcareous skeleton secreted externally; (b) alcyonarian coral, skeleton laid down by ectodermal cells in the mesoglea; (c), (d) 'true' corals with massive external skeletons.

often on top of the skeletons of their dead precursors because of past changes in sea level. Coral skeletons may extend for almost a kilometre and a half in depth, giving us a record of changes in sea level back to the Cretaceous era. Changes in sea level have enabled fringing reefs, barrier reefs and atolls (coral islands) to be formed.

Coral reefs have been called the 'rain forests of the sea' because they support richer ecosystems than any other marine habitat. Reefs provide surfaces for the growth of sedentary organisms and shelter for mobile animals. They also provide food: work on the coral *Acropora* on the Australian Great Barrier Reef shows that up to half of the carbon production is exuded as mucus. This not only protects coral polyps but also traps food particles that supply the whole coral ecosystem. Corals are inhabited by a great variety of fish, crustaceans, echinoderms and many other invertebrates: more than 93 000 coral-dwelling species have so far been described.

Like rain forests, coral reefs are sadly being destroyed, partly by predation, e.g. by periodic population explosions of the crown of thorns starfish, but mainly by ourselves. Damaging human activities include pollution of the sea, nutrient enrichment due to agriculture, fishing by dynamite, rock mining and removing corals for sale. Corals near the coast are smothered by sediment from the land, following deforestation and coastal building development. Too frequently nowadays divers find that corals have been 'bleached', stripped of their algae, leaving only white skeletons devoid of life. An example of long-term decline comes from the Caribbean, with a reduction of living coral by 80% over three decades.

Recently, climate change resulting in sea temperatures too high for the algal symbionts has been identified as the main cause of bleaching. The coral's weakened resistance to disease-bearing bacteria, due to physiological stress, does further damage. Some Caribbean corals are changing their symbiotic algae to new strains with greater thermal tolerance, and corals are more abundant where this has occurred, but adaptation is slow and cannot keep pace with the rate of warming of the sea. A more recently recognised further hazard is acidification of the oceans by CO_2, causing calcareous skeletons to dissolve. Twenty-five per cent of the world's reefs have already been lost or degraded due to global warming, which at least in part is caused by human activity. Major conservation efforts are urgently required to maintain this most valuable marine ecosystem.

4.6 How are Cnidaria related to each other and to other phyla?

All four classes of Cnidaria can be traced back to Precambrian fossils, but their relationships are controversial. Past candidates to be the most primitive group of Cnidaria have included hydrozoan medusae, hydrozoan polyps, anthozoan polyps and the transient polyp stages of scyphozoans. Cladistic analysis based on morphology cannot give us convincing results, because the characters on which it can be based are few and variable. Molecular evidence from both large and small subunits of ribosomal DNA points to Anthozoa as primitive. Mitochondrial DNA provides one important piece of evidence: in anthozoans as in all other known Metazoa the mitochondrial DNA is circular (as in bacterial DNA) but analysis of 25 hydrozoans, 5 scyphozoans and 1 cubozoan revealed that they all have linear mitochondrial DNA, which is very likely to be a derived feature. This supports, but cannot alone prove, the morphological arguments suggesting that Anthozoa are nearest to the primitive Cnidaria.

Molecular evidence supports the view that multicellularity arose only once in evolution, with Porifera, Cnidaria and triploblastic animals as three distinctive multicellular lines. Attempts to derive triploblastic worms from one group of cnidarians rather than another have never been convincing. That any animal could theoretically be derived from a planula larva could merely be a tribute to the planula's total lack of distinctive structure, but such a theory is back in fashion (see Chapter 20).

Cnidaria make their living differently from other animals. Their basic simplicity of structure limits their behaviour but has allowed the evolution of great diversity, and, especially as corals, they may dominate the marine environment. These often very beautiful animals have many unique features, and we can look at them with fresh eyes.

Chapter 5

On being a worm

Any soft-bodied legless animal whose length exceeds its width is liable to be described as a worm, and many invertebrates fit this description. Four of the main worm-like phyla are discussed separately in later chapters, but there are many other different worms, belonging to phyla often castigated as 'minor', usually because they have a small number of species or are very small animals. This chapter introduces the variety of worms, after considering why worms should have evolved so many times and what muscular machinery is necessary for their locomotion.

5.1   Why are there so many different kinds of worm?

Mechanical facts about the molecules that make up animals mean that worms are very easily produced. Cells secrete extracellular compounds with charged molecular backbones: like charges repel, causing linear extension, and linkage between these large molecules provides orientation in a structure that will be anisotropic (i.e. have different properties in different directions). If a blob of soft tissue has such orientated fibres, any event such as growth or motion or external pressure will automatically turn that blob into a cylinder. Orientated fibres will guide and limit the direction of growth, and enable it to change its shape. No further genetic instruction is needed to make a worm, in its simplest form.

   A worm, then, is easily produced: why should such a structure be favoured by natural selection? Soft tissues are extraordinarily resistant, and damage is readily repaired. There is a large range of possible sizes, from less than a millimetre (as parasites or in the marine 'interstitial' habitat in between sand grains) to over 30 metres in the sea. Movement in one direction will be favoured by worm-like shape: an anterior end becomes established, usually with at least a simple 'brain' and sense organs, and the distinction between dorsal (top) and ventral (underside) surfaces confers bilateral rather than radial symmetry. Metazoa with this construction are commonly grouped as 'Bilateria'.

Worms are 'triploblastic': they have a third tissue layer, the mesoderm, from which their muscle fibres are derived. Mesoderm is most simply described as any cellular tissue occurring between the ectoderm and endoderm, becoming distinct from these layers at some point in early development. Both layers contribute to mesoderm formation in most triploblasts, but some primitive phyla (for example Acoelomorpha, Ctenophora) derive mesoderm from the endoderm only. The presence of mesoderm is an important difference from Cnidaria because it separates muscle fibres from cells of the other two layers and allows them to run in all directions. Although very small worms can be propelled by cilia, most move by muscle contraction (see Box 5.1).

5.2 How can muscles move a worm?

5.2.1 Hydrostatic skeletons

Water is incompressible: a closed bag of fluid can change its shape but not its volume. Soft-bodied animals lacking hard skeletons use this fact to re-extend their muscles, mostly using a fluid-filled body cavity. The earthworm is a familiar example: longitudinal and circular muscles contract alternately acting against the internal fluid, bristles grip the substrate and the worm moves forward by the process called 'peristalsis' (Figure 5.1a). Other examples of the use of hydrostatic skeletons in invertebrates include:

Burrowing (many worms). Circular contraction extends the anterior end, which takes a hold, then longitudinal contraction pulls up the posterior part.

Wave motion (nematodes, polychaete annelids and many others). Waves of contraction are propagated alternately along the sides of the worm, producing an 'S' shape, with backward pressure on the environment so that the worm moves forward (Figure 5.1b,c).

Jet propulsion (octopus, squid and cuttlefish and also some jellyfish). Although the coelenteron is an enclosed part of the outside world rather than a body cavity, it can still act as a hydrostatic skeleton.

Muscular waves (platyhelminths, the foot of snails) effect slow propulsion.

Parts of animals (spiders' legs, starfish tube feet) may work by muscles squeezing fluid, even though a hard skeleton exists.

5.2.2 Body cavities: different kinds

Some worms have no body cavity: they are described as 'acoelomate', and all that the muscles have to work against is the fluid pressure of water held inside cells or in spaces in the soft tissues. Muscular waves may be used to effect slow propulsion. Worms with no body cavities

Box 5.1 | Muscle

Muscles contract and relax: they cannot actively stretch. Therefore every animal must be able to re-extend its muscles before they can contract again. Much of animal design depends on this simple fact. Muscles often occur in antagonistic pairs (as in the muscles either side of our limb joints), where contraction of one extends the other. To have any effect, muscles must work against resistance in the form of a skeleton, which may be hard (as in ourselves and arthropods) or hydrostatic, based on the incompressibility of water. Invertebrates have evolved a great range of musculature within the constraints of these basic requirements.

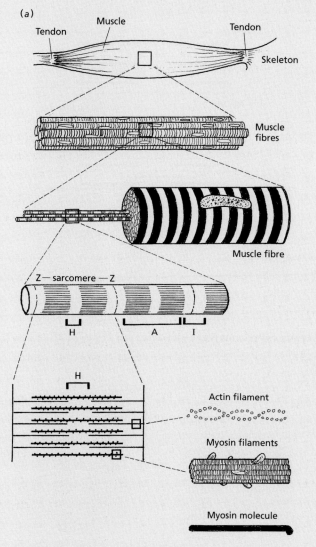

(a)

Tendon Muscle Tendon

Skeleton

Muscle fibres

Muscle fibre

Z — sarcomere — Z

H   A   I

H

Actin filament

Myosin filaments

Myosin molecule

(a) The hierarchy of skeletal muscle organisation.

Muscle fibres contract by shortening and/or by building up tension: some muscles ('isotonic') change mainly in length, while others ('isometric') change little in length but greatly in tension. All contraction depends on the sliding of microscopic filaments within the fibres (see diagram). Thin actin filaments slide between thick myosin filaments with formation of cross-bridges between them, and at the same time energy is released to power the contraction. The tension developed is proportional to the number of cross-bridges. This universal machinery also makes muscle elastic, to a varying extent.

(b) Shortening of muscle by sliding filaments: the unit which shortens is the sarcomere, between two 'Z lines'. Myosin (thick) filaments are shown dark. The muscle appears striated because myosin-containing regions (the 'A' bands, including the 'H' zones) alternate with regions containing actin filaments only (the 'I' bands).

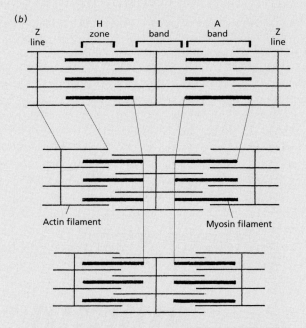

(b)

Actin filament   Myosin filament

The following kinds of muscle can be distinguished in invertebrates:

Striated muscle appears under the microscope to be cross-striped because the filaments are closely packed with their darker regions (where actin and myosin overlap, see diagram) coinciding. This fast-contracting, elastic, often isometric muscle is found all through the animal kingdom, from pulsating jellyfish bells to vertebrate voluntary muscle.

Insect asynchronous flight muscle is a special case of striated muscle, being the fastest and most isometric muscle known. Contraction is faster than the nerve input, enabling some small insects to beat their wings over 1000 times per second (see Chapter 15).

Helical smooth muscle may appear plain or may show stripes in a spiral round the fibre if the bands coincide. This is the commonest invertebrate muscle. A variant is 'long-fibred' smooth muscle.

Paramyosin muscle has very long and large filaments, as in the adductors of clams. Slow to react, very isotonic and inelastic, it can exert very much more force and sustain it for longer than any other muscle. These properties enable it to maintain tension at very different degrees of extension, as in the smooth muscle confined to vertebrates.

include Platyhelminthes and Acoelomorpha (Chapter 6), Nemertea (Chapter 7), Gnathostomulida and Mesozoa.

A much more effective hydrostatic skeleton is provided by a fluid-containing body cavity. Traditionally, animals with body cavities were divided into coelomates, pseudocoelomates and animals with haemocoels, on the basis of the following characteristics:

Coeloms are body cavities bounded on all sides by mesoderm. They occur in Annelida (including Pogonophora and Echiura), Sipuncula, Phorona, Brachiopoda, Bryozoa, Chaetognatha, Echinodermata, Hemichordata, Chordata and, although not as the main body cavity, in Mollusca and Arthropoda. Here already the traditional categories are open to argument, according to more precise definitions of a coelom.

'**Pseudocoels**' are not a single type of cavity: here the traditional classification breaks down. They are united only in lacking a mesoderm layer between the cavity and the gut. They may be derived from vacuoles within cells or by persistence of the first cavity to appear during development, the blastocoel (see Box 5.2). 'Pseudocoels' were said to occur in Nematoda (Chapter 8) and a variety of worms including Gastrotricha, Nematomorpha, Rotifera, Acanthocephala, Loricifera, Kinorhyncha and perhaps Priapula.

Haemocoels are persistent blastocoels (the first-formed cavity) expanded and filled with blood, acting both as body cavities and as substitutes for canalised blood systems. Haemocoels occur in Mollusca (Chapter 10) and in Arthropoda (Chapter 12). With a haemocoel rapid blood circulation is difficult to achieve, but all the tissues are bathed in blood.

These body cavities are of great functional importance, primarily because they serve as hydrostatic skeletons. It is not surprising that such a useful feature has evolved many times separately. We now realise that the presence or absence of a particular kind of body cavity is no guide to phylogeny. For example, all animals possessing a coelom may be described as 'coelomates', but we now know that there is no close evolutionary relationship between all coelomates. The term 'pseudocoelom' has been abandoned.

5.2.3 Lattices

All hydrostatic skeletons need the support of connective tissue lattices: threads of collagen, incompressible in length, are wound helically round the soft body (Figure 5.1d). The animal can therefore contract evenly without bulging or kinking as the lattice angle changes. The lattice is a spirally coiled fibre like a spring, which on stretching becomes thinner, i.e. the lattice angle becomes smaller, and on compression becomes fatter, increasing the lattice angle. At one extreme the lattice angle tends to zero, at the other to 90 degrees, but in an animal the stiffness of the bounding layer prevents the angle reaching either of these extremes.

Graphical representation of lattice action considers the worm as a cylinder. Then

$$\text{volume} = \text{length} \times \text{cross-sectional (TS) area}$$

and this area is proportional to the square of the diameter. Since the fibre of which the lattice is made cannot change in length, the volume and the lattice angle will be related as shown in Figure 5.1e. Since a worm does not change in volume but in length, it must be represented on this graph by a horizontal line such as *AB*. *A* and *B* represent extremes of the lattice system: the worm cannot be

(a)

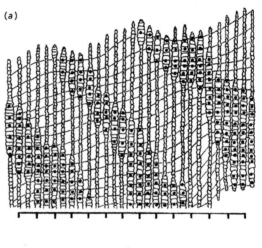

(b)

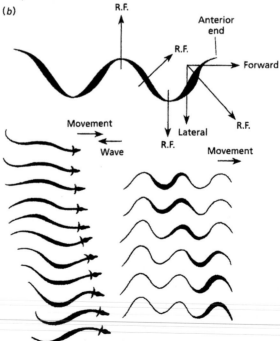

Fig. 5.1 Worm locomotion. (a) Peristalsis: successive stages in the forward movement of an earthworm. (b) Waves of muscle contraction passing from the anterior to the posterior end of an undulating worm: forces generated (above) and resulting locomotion (below).

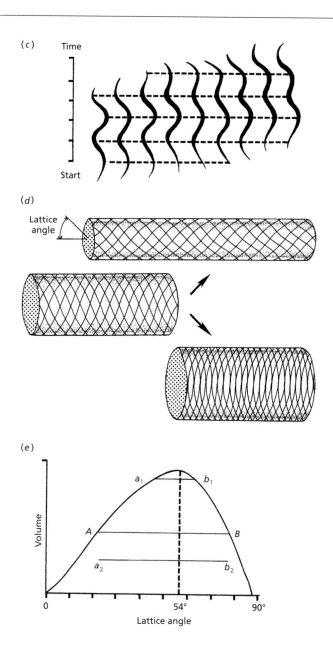

(c)

Time

Start

(d)

Lattice angle

(e)

Volume

Lattice angle

0   54°   90°

a_1 b_1

A   B

a_2 b_2

Fig. 5.1 (contd.) (c) The nematode *Haemonchus contortus* creeping over the surface of an agar gel: tracings from successive photographs at 0.3 second intervals. (d) Connective tissue lattices, showing how the spirally arranged collagenous fibres control the effect of muscle contraction. (e) Graph to show the relationship between volume and lattice angle in a cylinder. *AB* represents most worms, $a_1 b_1$ any terrestrial planarian or nemertine, $a_2 b_2$, *Lineus longissimus*, a swimming nemertine.

longer and thinner than it is at *A*, nor shorter and fatter than at *B*. The line of the graph defines the only positions where the lattice allows the worm to be circular in cross-section: the worm must therefore flatten (becoming oval in TS) as it changes between *A* and *B*. Different worms will have different positions of *AB* on the graph; a land-living worm, for example, will tend to be circular in cross-section since this shape presents a minimal area for water evaporation, while some swimming worms can become flattened (Figure 5.1e).

5.2.4 The disadvantages of a hydrostatic skeleton

A hydrostatic skeleton has the great advantage of being resistant to impact damage: worms do not readily break, buckle or burst, distortions are not harmful and considerable changes in shape are allowed. There are, however, formidable disadvantages. Firstly, the skeleton depends upon hydration. To be amply supplied with water the animal must at least be physiologically aquatic (i.e. confined to damp places if terrestrial). The weight of fluid required offsets the advantage of not having a heavy hard skeleton (the threads of the lattice are so strong that they can be very thin and light). Secondly, to move the whole body there must be a great deal of muscle, unlike animals with legs for leverage where muscle can be concentrated for efficiency. Further, in a crawling worm there will be a large surface impeded by friction on the substrate. Locomotion is even more uneconomical in energy requirements because the whole body must be accelerated and decelerated all the time. Thirdly, such a system is much harder for nerves to control, and impossible to control precisely: a beetle (for example) can know exactly where its legs are and the degree of bending at each joint, but no such information exists for a worm. Quite apart from the vulnerability to predators of a soft body, the advantages of being a worm should not be overstated.

All the same, the range of worm phyla is impressive.

5.3 What worm phyla are known?

The four main phyla are described in Chapter 6 (**Platyhelminthes**), Chapter 7 (**Nemertea**), Chapter 8 (**Nematoda**) and Chapter 9 (**Annelida**). This chapter introduces the smaller phyla, illustrated in Figures 5.2 to 5.4, with brief text indicating for each phylum the number of known species (very approximately), the size of the animals, their habitat, and distinguishing features.

Why include these 'minor phyla' in this introduction to invertebrates? Partly to show the range of body plans occurring in 'worms', partly on account of their great relevance to our consideration of phylogenetic relationships. In the past, a number of these phyla (including nematodes) were united as 'pseudocoelomates' in a group called 'Aschelminthes'. Morphological investigation using electron microscopy abolished this group a number of years ago and more recently molecular and morphological evidence has been combined to establish a very different grouping of all these worms (see Chapter 20). As has been explained, body cavities are not useful guides to phylogeny. The terms 'acoelomate' and 'coelomate' are retained as descriptive; the term 'pseudocoelomate' is too imprecise to be retained at all.

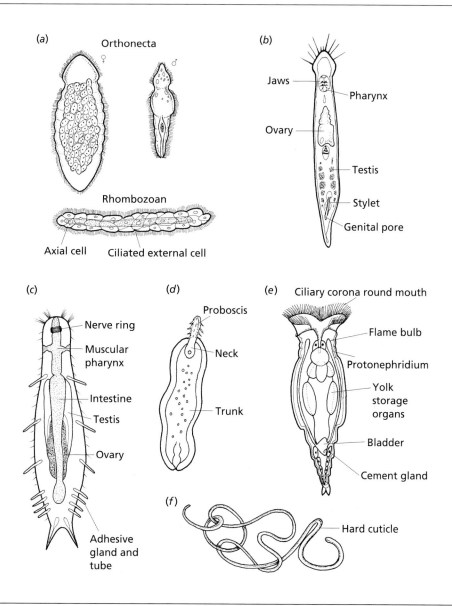

Fig. 5.2 The diversity of worms: (a) Mesozoa from the two subgroups: orthonectans, female and male, and a rhombozoan (dicyemid); (b) a gnathostomulidan; (c) dorsal view of a gastrotrich, seen as if transparent; (d) an acanthocephalan; (e) a bdelloid rotifer; (f) a nematomorph.

The smaller phyla of what can be called 'worms' are:

Mesozoa (Figure 5.2a). About 50 species. Length about 0.5 mm, marine. Endoparasites with complex life cycles and extremely simple structure: an outer ciliated layer, very few non-reproductive cells in the inner (mesoderm) layer, no body cavity and no endoderm. The phylum includes two different groups, the Rhombozoa (dicyemids) and the Orthonecta.

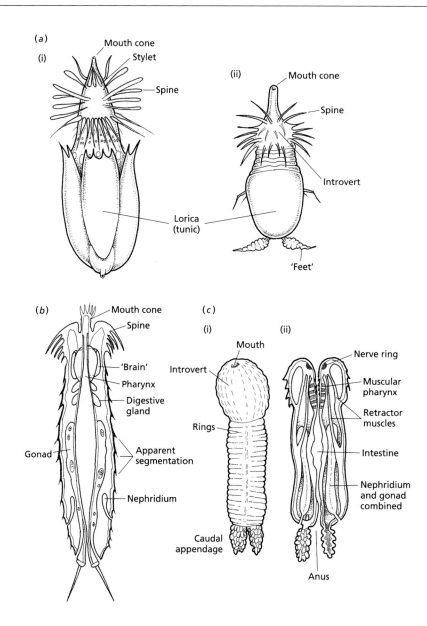

Fig. 5.3 The diversity of worms (continued): (a) dorsal view of (i) adult and (ii) larval loriciferan; (b) a kinorhynch, showing the internal anatomy; (c) *Priapulus*, showing (i) the external morphology and (ii) the internal anatomy.

Myxozoa. Another group of endoparasitic worms recently recognised as a separate phylum (see Chapter 20).

Gnathostomulida (Figure 5.2b). At least 100 species. Length <1 to 4 mm, marine. Acoelomate miniature anaerobic flatworms gliding through the sand, often very numerous in the interstitial habitat. Their name derives from the specialised muscular pharynx with jaws that grasp food.

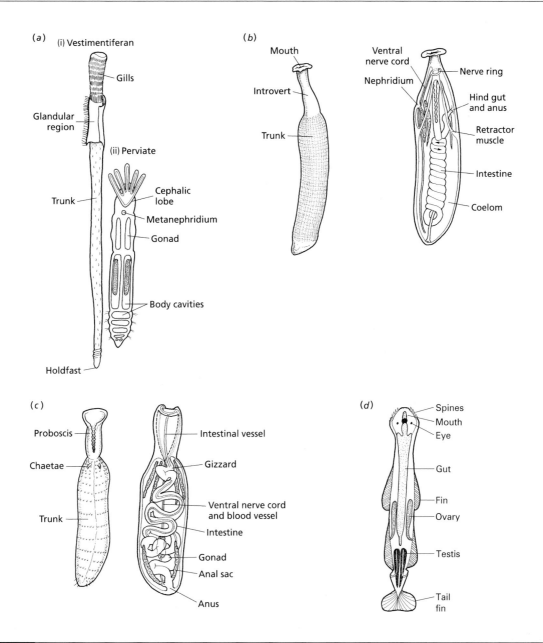

The diversity of worms (continued). Some worms with coeloms: (a) Pogonophora (Siboglinidae): (i) a vestimentiferan and (ii) a perviate; (b) a sipunculan, external view and internal anatomy as shown by dissection from the left side; (c) an echiuran, external view and internal anatomy shown by dissection from the dorsal side; (d) external view of the chaetognath *Sagitta elegans*. Pogonophora (Siboglinidae) and Echiura are now usually included in the phylum Annelida.

Micrognathozoa. A single species from a cold spring in Greenland was described in 2000. These are very small animals resembling Gnathostomulida but characterised by a unique jaw structure made from rods of cuticle. Two rows of multi-ciliated cells provide locomotion.

Gastrotricha (Figure 5.2c). At least 450 species. Length <0.5 mm, marine. Like the Gnathostomulida, they glide through often anoxic sand or mud. They have external cilia and a protective cuticle, not strengthened as in nematodes but with adhesive tubules. The cilia sweep bacteria and other food into the mouth.

Acanthocephala (Figure 5.2d). About 1000 species. Length up to 800 mm. Specialised endoparasites, the larvae live in insects and crustaceans and the adults in the gut of vertebrates. There is a proboscis covered in hooks, a body cavity and no gut. The surface cuticle has conducting channels.

Rotifera (Figure 5.2e). At least 2000 species. Length up to 3 mm. Most live in fresh water. They are propelled by a crowning 'wheel' of cilia, the corona. The ciliated epidermis has an intracellular 'lorica' that can form a resistant resting stage in unfavourable conditions: this, coupled with remarkably rapid reproduction in favourable conditions, has made the phylum very successful. Parthenogenesis is common; the class Bdelloidea never has any males and sexual reproduction is unknown.

Nematomorpha (Figure 5.2f). At least 250 species of 'hair worms'. May be 360 mm long and only 1 mm in diameter. They live in fresh water or damp soil, with one marine genus. Like nematodes, they have a thick outer cuticle, a body cavity and only longitudinal muscle, contracting in propagated waves for locomotion. The young are parasitic in arthropods and the adults of some species do not feed.

Loricifera (Figure 5.3a). More than 100 species. Length <0.5 mm, marine. Discovered in 1983, these marine interstitial burrowers have an introvert (eversible head) and a trunk, with a body cavity, encased in a 'lorica'. The larvae have 'feet' posteriorly.

Kinorhyncha (Figure 5.3b). At least 150 species. Length <1 mm, marine. These very small carnivorous worms have a body cavity and an introvert used for burrowing in mud. They are remarkable in being thoroughly subdivided into 13 parts, a subdivision resembling the 'segmentation' found in annelid worms (see Chapter 9), but kinorhynchs are not otherwise similar to any segmented animals.

Priapula (Figure 5.3c). At least 16 living species, 11 fossil. Length up to 60 mm. Like kinorhynchs, priapulans are carnivores burrowing in marine mud; here the introvert is large and conspicuous. The body is covered in a chitinous cuticle and the body cavity is probably a true coelom. They have long been seen as an ancient, isolated phylum.

Pogonophora (Siboglinidae) (Figure 5.4a), marine, are most unusual in that they have no alimentary canal (gut). They are segmented coelomate worms secreting tubes of chitin and protein in which they live, sticking upright from the ocean floor. Tissue replacing the gut is packed with sulphur bacteria that are the main source of food, but also the outer surface of the worm absorbs dissolved organic matter. The two subgroups are:

> **Perviata**. At least 95 species. Length of tubes up to 8000 mm. They occur in soft sediments to a depth of 9000 metres. They need reducing sediments, which supply reduced sulphur compounds to their bacteria, overlain by oxygenated water supplying oxygen to the worm's haemoglobin.

> **Vestimentifera**. '15 species', but the growing exploration of ocean depths is rapidly increasing this number. Tubes may be 2 metres long and 30 mm in diameter. They are plentiful in the deep ocean vents where all life is based on sulphur bacteria. The trunk below the tentacles (gills) bears wing-like 'vestiments' in place of the bristles occurring in Perviata. Some juvenile vestimentiferans have recently been found to have guts, lost in the adult.

Since their discovery about 100 years ago, Pogonophora have been regarded as a separate phylum. It has now been suggested (with support from molecular evidence) that they are a much modified group of polychaetes. 'Siboglinidae' is an older name for the group, now increasingly adopted, to place them as a family of annelids rather than as a separate phylum.

Sipuncula (Figure 5.4b). At least 320 species. Length 1−300 mm, marine. They are sedentary deposit feeders, having tentacles, a retractable introvert and a trunk. Unsegmented coelomates unlike annelids in appearance.

Echiura (Figure 5.4c). At least 150 species. Length up to 500 mm, marine or brackish. Sedentary deposit feeders known as 'spoon worms', as sediment is collected by the elongated proboscis projecting from the trunk. Unsegmented coelomates. Echiura are no longer seen as a separate phylum but as polychaete annelids that have lost their segmentation. Recent discovery of the segmental organisation of the nervous system in the larva supports the view that Echiura are not primitively unsegmented but have lost segmentation during evolution.

Chaetognatha (Figure 5.4d). At least 90 species. Length 3–120 mm, marine. Chaetognaths are carnivores, major predators in the marine plankton. These 'arrow worms' are streamlined, with rigid transparent bodies and horizontally projecting fins. They swim very fast and pounce on their prey, seizing it with the two groups of movable spines projecting on either side of the mouth. The worms are coelomate, with longitudinal and oblique muscles contracting against the hydrostatic skeleton, a well-developed nervous system and haemal sinuses in place of a circulatory system. They are hermaphrodites with direct development. Chaetognaths are a small and uniform group, but often very numerous as individuals. In the English Channel two species are very common, *Sagitta elegans* from the Atlantic and *S. setosa*, which is an indicator species for coastal Channel water. At Plymouth in Devon an increased proportion of *S. elegans* indicates an influx of Atlantic water, overlying the Channel water, because this species has gas-filled pockets in the intestine causing the worm to float higher in the sea. The relationships of this isolated phylum have long been a puzzle, as their development shows some deuterostome features but other evidence suggests that they are protostomes. Study of the mitochondrial genome is resolving this puzzle, as will be explained in Chapter 20.

Box 5.2 | Protostomes and deuterostomes

Chapter 19 gives an account of development, but an introduction is needed sooner, because early development may be a key to evolutionary affinities as well as to adult form (the word 'development' is applied to individual change, **ontogeny**, and not to evolutionary change, **phylogeny**). Developmental patterns depend in part on the quantity of yolk in the egg: yolk is invaluable to feed the embryo but it is heavy and inert, impeding the division and movement of cells in which it is contained. The following account applies primarily to animals (such as most marine invertebrates) where the eggs contain little yolk and usually hatch early into feeding larvae.

According to alternative methods of development at three early stages, all phyla with three cell layers (triploblasts) may be divided into two large groups:

Protostomes: all the triploblast phyla, except
Deuterostomes: echinoderms, chordates and possibly a few others.

The three developmental stages which distinguish protostomes and deuterostomes are (1) cleavage, the first division of the fertilised egg; (2) the fate of the blastopore, the earliest opening of the future gut; (3) the formation of mesoderm and (in animals which have one) the coelom.

Cleavage

A non-yolky fertilised egg typically divides into two cells and then each divides at right angles into two cells lying in the same plane. The third division may be either radial (deuterostomes), with cells directly on top of each other, or spiral (protostomes), where cells are placed between those underlying them (see diagram). Cleavage continues, and in either case produces a ball of cells or blastula, hollow because the cells secrete fluid into the centre, forming the blastocoel or primary body cavity.

The blastopore

Next, cell movements called gastrulation produce a two-layered structure, where an inner endoderm approaches the outer ectoderm, thereby much reducing the blastocoel. The endoderm-enclosed cavity is called the 'archenteron' or future gut, opening by a hole called (rather misleadingly) the blastopore. This hole becomes the mouth in protostomes and the anus in deuterostomes (so-called because the mouth is a secondary opening, elsewhere).

Mesoderm

In protostomes the future mesoderm is determined very early in cleavage and is identified in the early larva as solid blocks of cells either side of the posterior end of the gut. These cells then split or hollow out to form a 'schizocoelic' coelom. The diagram illustrates these processes both for an annelid (segmented; the coelom is enlarged and almost obliterates the blastocoel) and for a mollusc (unsegmented; the coelom is very small, leaving a large persistent blastocoel). In deuterostomes mesoderm is formed by out-pouching of sheets of cells in the gut wall; the pouches separate, each enclosing a part of the archenteron, which becomes an 'enterocoelic' coelom. In summary:

Developmental stage	Protostome	Deuterostome
Cleavage	Spiral	Radial
Fate of blastopore	Mouth	Anus
Mesoderm and coelom	Early determined schizocoelic	Gut out-pouching enterocoelic

That three distinct characters in early development unite to distinguish protostomes and deuterostomes is an indication of evolutionary relationships. The division is generally accepted, and supported by molecular evidence. The categories are not, however, as clear and absolute as the above formulation suggests: there are examples of radial cleavage combined with a protostome mouth, and anomalous species can be found in most phyla. Again and again in biology, there is a clear framework but exceptions abound, because natural selection is operating.

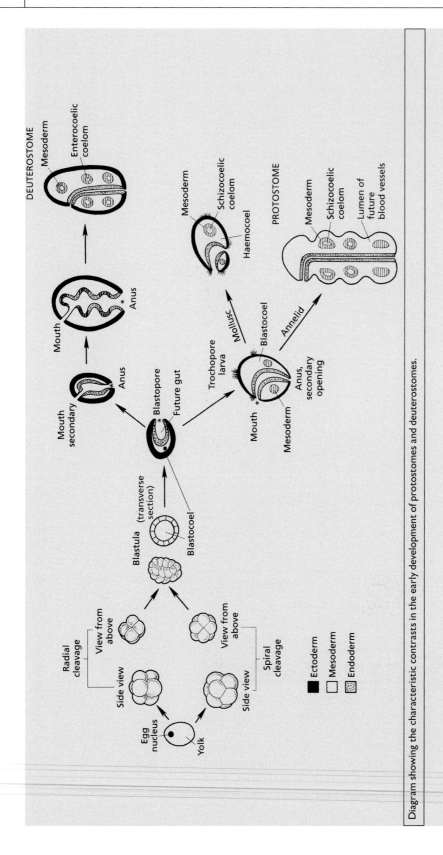

Diagram showing the characteristic contrasts in the early development of protostomes and deuterostomes.

5.4 Do Ctenophora belong among the worms?

On shape, symmetry and method of locomotion the answer is clearly that ctenophores are not worms, but this isolated small phylum has features that may further define a worm by contrast.

Ctenophora (Figure 5.5). At least 100 species, marine. Ctenophores are widespread and abundant (as individuals rather than species) in the oceans, with new deep-sea forms currently being discovered. Although they are superficially like medusae, being (often beautiful and delicate) bags of jelly floating and swimming in the sea, closer inspection shows that they do not closely resemble Cnidaria (traditionally their companion 'coelenterates'). Above all, they have no stinging cells. Nor are they radially symmetrical, since all but one small group bear a pair of tentacles and digestive canals. They are 'biradial' with no front end, in contrast to a bilaterally symmetrical worm with an anteroposterior axis and a brain in front. And they are triploblastic: mesoderm is present, forming a network of large muscle cells separate from the inner and outer layers. Clearly, animals without stinging cells or radial symmetry and with mesoderm cannot be classed with Cnidaria.

Ctenophores swim by means of large cilia fused into 'comb plates' arranged transversely across each of the eight 'comb rows', coordinated by the apical sense organ from which they radiate (Figure 5.5). This unique machinery enables ctenophores to swim more efficiently than small animals using normal cilia, although they lack the muscular development of a worm. Ctenophores are able to be pelagic predators because they have unique adhesive cells, called 'colloblasts' or lasso cells (Figure 5.5), but these are not at all like

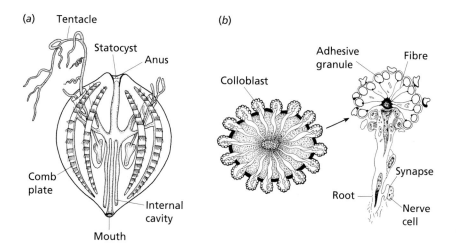

(a) Tentacle (b)

Statocyst
Anus
Adhesive granule
Fibre
Colloblast
Comb plate
Internal cavity
Synapse
Root
Nerve cell
Mouth

Fig. 5.5 Ctenophora: (a) lateral view of a typical ctenophore; (b) section through a tentacle of a ctenophore, with detail of a colloblast.

cnidarian nematocysts. The mouth is used for ejection of faeces, but there is a separate anus for removal of metabolic waste.

Altogether, ctenophores are unusual animals, hard to place. The mesoderm is unlike that of all other triploblastic animals (except acoelomorphs) in being purely endodermal in origin and almost all becoming muscle. They have unique cell types: ciliated comb cells, colloblasts, giant unstriped muscle cells; electron microscopy also reveals unique cell junctions. Their isolation is further emphasised by their early development: the first cell divisions form a unique flat plate rather than a ball, and at this stage the fate of each cell is determined by its lineage rather than by interactions with other cells (much as in nematodes: see Chapter 8). Future ectoderm cells stay in clusters, the other cells disperse, and endoderm cells at the oral end of the embryo become mesoderm. The embryo develops directly into the adult in some species, but in others there is a unique globular ciliated larva.

Ctenophores do not resemble worms any more than they resemble Cnidaria, and, except possibly for the minute Placozoa (see Chapter 3), no other group of animals appears to be at all closely related to ctenophores.

Chapter 6

Platyhelminthes and Acoelomorpha

Platyhelminths and acoelomorphs are two of the simplest groups of animals with three cell layers and bilateral symmetry. When animals are simple in structure we tend to assume that they are primitive, meaning that these animals are representative of early ancestral forms, but this assumption is not necessarily justified: simple animals may have evolved from ancestors more complex than themselves (that is, the simplicity may be secondary), and this can apply to the group as a whole. Simplicity may not be a reliable guide to phylogeny.

Is the simplicity of platyhelminths and acoelomorphs a primitive characteristic or has it been secondarily derived? To answer this question, the following account first describes the body plan of each group and then discusses the main specialisations which have arisen in much the larger group, the platyhelminths. An assessment of relationships is then introduced, to be taken further in Chapter 20 when molecular evidence is discussed.

6.1   What is the body plan of the platyhelminths?

The basic structure of platyhelminths is very simple (Figure 6.1). They are 'triploblastic', i.e. have three cell layers, outer ectoderm, mesoderm and inner endoderm. In the free-living forms there is an elaborate hermaphrodite reproductive system and an excretory system of ectodermal tubes, the protonephridia (flame cells). Free-living platyhelminths have a head end with anterior sense organs and a rudimentary brain. They move by means of the cilia or by contraction waves in muscle working against fluid pressure of the body contents (gut, parenchyma, genitalia etc.).

These animals are commonly regarded as primitive on account of a number of negative characters: the mouth is the only opening to the gut, and they have no body cavity, no respiratory system, no blood system, no appendages and no hard skeleton. Animals without a respiratory or a blood system are necessarily rather

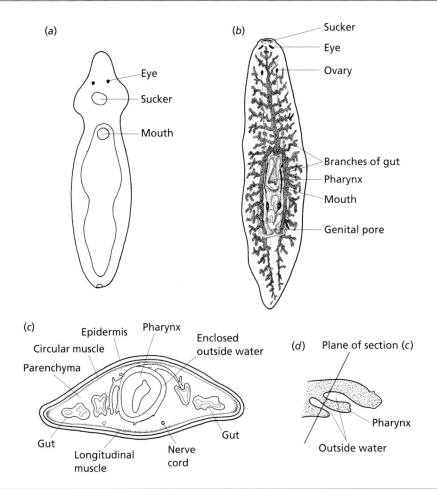

(a)

Eye
Sucker
Mouth

(b)

Sucker
Eye
Ovary

Branches of gut
Pharynx
Mouth

Genital pore

(c)

Epidermis   Pharynx
Circular muscle
Parenchyma

Enclosed
outside water

Gut

Longitudinal
muscle

Nerve
cord

Gut

(d)   Plane of section (c)

Pharynx

Outside water

Fig. 6.1 Platyhelminths: the structure of (a) a catenulid, (b) a planarian (longitudinal section) showing branching gut (both species about 10 mm in length); (c) transverse section of planarian in the pharyngeal region; (d) side view of head and proboscis.

flat, being limited by the diffusion of oxygen, and since they are soft-bodied with the length greater than the width, they are known as flatworms.

6.2   What groups of worms constitute the Platyhelminthes?

Free-living platyhelminths with ciliated epidermal cells. They are the **Turbellaria**, a heterogeneous group including the Catenulida (Figure 6.1a), small mostly freshwater worms with many simple characters, and the Rhabditophora, a large and varied group including planarians (Figure 6.1b,c), in fresh water or in damp earth, and marine polyclads.

Parasitic platyhelminths. Most platyhelminth species are parasitic (see Box 6.1). There are three classes:

Monogenea, flattened 'flukes', ectoparasitic on aquatic animals (mostly fish)

Trematoda, including **Digenea** (e.g. the liver fluke), endoparasitic in vertebrates

Cestoidea, tapeworms, highly modified endoparasites, widespread in occurrence.

In all these parasitic forms, the epidermis is replaced by a 'tegument' protecting the parasite and governing what it can absorb from its host.

6.3   What are the Acoelomorpha?

The acoelomorphs comprise the Acoela and the Nemertodermatida.

6.3.1   The Acoela

The Acoela (Figure 6.2a,b) are very small (about 3 mm long) and simple, even when compared to platyhelminths. They have no hollow gut, but the innermost endodermal tissue is a digestive layer that may form only after feeding. This tissue may be 'syncitial', with nuclei scattered in the cytoplasm and no cell boundaries. Mesoderm in the form of muscle cells arises from the endoderm cells; unlike platyhelminths there is no ectodermal source of mesoderm. Acoela have no protonephridia or other excretory structures. The eggs and sperm are not contained in gonads. Coordination is achieved by a nerve network with few distinct nerve cords and minimal (and chemically unusual) brain development, very unlike that of platyhelminths.

Are these animals primitively or secondarily simple? Reduction of a more specialised form was suggested by the discovery of many very unusual features in Acoela. The sperm structure is unique (see below), the cilia and their rootlets are highly unusual and there is hardly any intercellular substance. There is no larval stage. Most tellingly, in earliest development spiral cleavage is quite different from that of other spirally cleaving animals (see Chapter 19). In platyhelminths, annelids, molluscs and others the second and each subsequent division places the division products over the furrows between the underlying cells, forming a spiral (Figure 19.1) which turns alternately clockwise and anticlockwise as division proceeds. In Acoela two separate spirals are formed, starting at the second division, producing pairs, i.e. 'duets' rather than quartets of cells. Subsequent clockwise and anticlockwise turns do not alternate but have a different distribution. The future fates of the early cleavage products is also different, and far more dependent on reactions

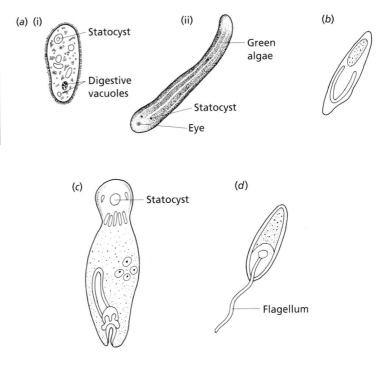

Fig. 6.2 Acoelomorphs: the structure of (a) the acoelan *Convoluta roscoffensis* (about 3 mm in length), (i) condensed and (ii) extended to show green algae; (b) acoelan sperm; (c) a nemertodermatid (about 10 mm in length); (d) nemertodermatid uniflagellate sperm.

between cells than in the typical very stereotyped spiral cleavage. These differences have been known for many years but were ignored until molecular evidence (discussed in Chapter 20) suggested that Acoela were not only primitively simple but also altogether separate from Platyhelminthes.

6.3.2 Nemertodermatida

These small worms (length about 10 mm) (Figure 6.2c) share most of the simple features of Acoela. The gut is not totally absent but almost occluded by endodermal cell processes, and there may or may not be a mouth. The sperm (Figure 6.2d) are unlike both acoelan sperm and the specialised sperm of most platyhelminths (Figures 6.2b and 6.3b). Nemertodermatids also differ from them all in having external fertilisation. Resemblances to Acoela include the specialisations of ciliary structure, formation of mesoderm from endodermal cells only and the absence of protonephridia. The brain and nervous system are only slightly more elaborate than in Acoela. Information about early development will clearly be very interesting: unpublished reports suggest that duets are formed, leading to bispirals, as in Acoela.

Molecular evidence (Chapter 20) now establishes that the Nemertodermatida belong with Acoela and are primitively simple, being direct descendants of primitive Bilateria. Acoela and Nemertodermatida are combined as 'Acoelomorpha', a phylum quite separate from the Platyhelminthes within which they were traditionally placed.

6.4   What is specialised about modern platyhelminths?

6.4.1   Reproduction

Ancestral triploblasts probably resembled other simple marine invertebrates in having separate sexes, shedding sperm and eggs (without much yolk) straight into the sea and hatching early as small feeding larvae. Most modern platyhelminths by contrast have most elaborate systems of muscular glandular organs forming extremely complicated hermaphrodite reproductive systems (Figure 6.3a). Most turbellarians live in fresh water, and as is typical in this habitat they have internal fertilisation of large yolky eggs, which develop directly without delicate larval stages. Parasitic forms have particular problems. They have even more elaborate reproductive systems and a series of larvae that facilitate transfer between hosts (see below).

Spermatozoa provide a striking example of specialisation in the Turbellaria. Most animals have the familiar 'tadpole-like' sperm moved by a single flagellum, but nearly all turbellarian spermatozoa have two flagella. The characteristic cap or 'acrosome' is absent, the nucleus is at the anterior end and the two flagella extend through the lateral cytoplasm of the cell (Figure 6.3b). There are exceptions: catenulid sperm are ovoid with no flagella but multiple short ciliary structures by which they move. Biflagellate sperm may be correlated with internal fertilisation and the need for the sperm to travel through tissue to reach the egg. Yet internal fertilisation is common among invertebrates and the platyhelminth biflagellated spermatozoa are unique among animals.

Another specialisation in most platyhelminths is that yolk is not produced within the egg cell but in separate cells, which are later incorporated into the egg capsule.

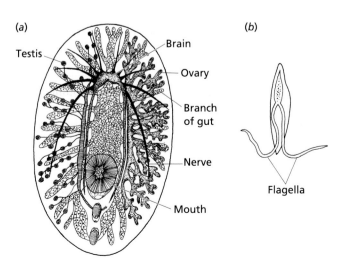

(a)

Testis
Brain
Ovary
Branch of gut
Nerve
Flagella
Mouth

(b)

Fig. 6.3 Reproduction in Turbellaria: (a) the hermaphroditic reproductive system of a polyclad turbellarian in ventral view, with testis shown only on the left and ovaries shown only on the right; (b) turbellarian biflagellate sperm.

The remarkable powers of regeneration of many planarians may also represent specialisation: as little as one-sixteenth of the body can reconstitute a new small animal. Also, in adverse conditions the animal may lose its differentiated structures and shrink, increasing again in size and complexity when the environment is favourable.

6.4.2 The epidermis

This is another very specialised structure in Turbellaria. The multiplicity of muscular gland organs is probably very significant. Glands produce complex forms of mucus that protect the animal and, in some way not understood, facilitate both locomotion and attachment to surfaces.

6.4.3 Other systems

In Turbellaria other systems also show great elaboration in detail. The nervous system is a variable mixture of specialised and unspecialised, advanced and primitive components, again with the greatest variation in the groups that are candidates to be the most primitive Turbellaria. There is a corresponding variety of simple epidermal sensory receptors. Muscle is not well developed, as would be expected in animals where there is so little provision for their re-extension after contraction, yet the ultrastructure of some of the faster muscle is reminiscent of striation (see Box 5.1).

6.4.4 Predation and parasitism

Turbellarians are nearly all predators, capturing their food with an eversible pharynx. The complexity of the gut increases in the larger turbellarians: with no transport system, the whole body can be supplied with food only by gut branches. The evolution of parasitism, a special form of predation, is the outstanding specialisation in platyhelminths.

Parasitism was a fruitful evolutionary direction for platyhelminths (see Box 6.1). Platyhelminths are relatively small and thin and inactive: without any means of transporting food and oxygen round the body they cannot become larger, and their muscle system does not allow much activity. Parasitism is a way of life where these characteristics are not disadvantageous. Energy can be concentrated on reproduction, and the complex hermaphrodite genitalia of free-living platyhelminths, coupled with internal fertilisation, can readily evolve the great fecundity required by parasites. There is a large number of parasitic species with a great variety of life cycles, very often with an intermediate host. For example in the liver fluke *Fasciola* (Figure 6.4a), where the adult lives in a sheep, the eggs are shed into damp grass where they hatch into free-swimming larvae (unlike the adult this stage has a ciliated epidermis, and simple eyes). These larvae enter the lung of a snail, where they form cysts that release new larvae which multiply asexually

Box 6.1 | Parasitism

Parasitism is a specialised form of predation. A parasite lives inside, or attached to the outer surface of, the body of a living host of another species, from which it obtains its food. The parasite benefits from the association at the host's expense; typically it harms but does not kill the host and thus become deprived of a home.

Parasites have evolved in almost every phylum but are particularly prevalent in protista (which are outside the scope of this book), platyhelminths (see Figure 6.4 for two examples of life cycles), nematodes, acanthocephalans and certain arthropods (see Chapter 13 for special attributes of parasitic crustaceans). Parasites are very widespread and likely to be important in any ecosystem. They are sometimes described as degenerate because they have lost some of the elaborations of their free-living relatives, but they are more accurately seen as specialised, being highly adapted to their particular way of life.

For example, internal parasites have little need for organs of locomotion and food capture, but have particular difficulty in completing their life cycle. The young cannot overcrowd the parental host, so they must find new hosts; many fail and then die, so that a large number of young must be produced. Typically there is an elaborate life cycle with a series of larvae (perhaps multiplying asexually) as agents of dispersal and host-finding. These larvae frequently find an intermediate host, perhaps one eaten by the primary host, to facilitate transfer.

The life of a metazoan parasite demands special adaptations: an ectoparasite must be able to attach to the host's outer surface, an endoparasite must penetrate the host, find and perhaps attach itself to a particular tissue and resist the host's defences. A gut parasite must resist the host's digestive enzymes, a blood parasite must adapt its surface antigens to resist the host's immune system. Such adaptations are not made once for all: there is a continuing 'arms race' between parasite and host. This demands continual individual variation on both sides, which can be supplied most rapidly by sexual reproduction. The ubiquity of parasitism and the demands of the arms race may indeed be a powerful reason for the continuation of the apparently wasteful procedures of sexual reproduction. Sudden changes in parasitic infection may in part be responsible for the often overwhelming success of species invading a new country, because the invaders do not necessarily bring their parasites with them. Such invasions are increasingly common in these days of rapid transport and may quickly decimate the native fauna.

Parasites may even alter the behaviour of their hosts to suit their own requirements: for example, trematode flukes parasitising certain snails cause the host to seek the light, bringing them to open places where predation by birds is much more likely. This brings the flukes to the birds, which are their next hosts, but does not at all benefit the snails. This behaviour change has been attributed to action by the parasite on the host hormones, but that a parasite may directly affect gene expression has been shown in another species of trematode-infested snail. At infection this snail suffers a three-fold increase in production of the messenger RNA which codes for a brain neuropeptide, causing the snail to cease all sexual activity and rapidly increase in size: it can then accommodate a larger number of the parasites.

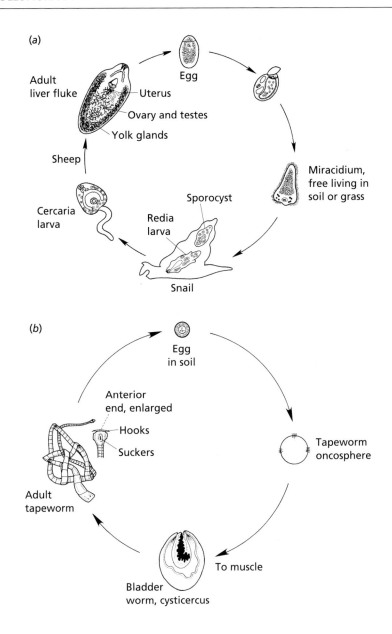

Fig. 6.4 The life cycles of two parasitic platyhelminths: (a) the liver fluke *Fasciola*, a trematode; (b) the tapeworm *Taenia*, a cestode.

to increase their numbers even more, and the final larvae wait in the grass. If they are eaten by sheep, the cycle is complete. Cestodes such as the human tapeworm *Taenia* (Figure 6.4b) are even more modified from their free-living ancestors. A 'head' fastened into the host intestine buds off a series of 'proglottides' that are little more than bags of reproductive organs; there is no gut, and food is absorbed all over the surface. Again there is a complex life cycle: eggs pass out into the soil where they develop into a hooked embryo. If eaten by a grazing animal, the embryo makes its way to the pig's muscle and encysts, ready to be eaten by a human.

6.5   How are platyhelminths related to each other?

Primitively, platyhelminths were free-living animals; parasitic adaptations are universally agreed to be secondary. The ultrastructure of the tegument is unique and is common to all three parasitic classes, strongly suggesting their joint origin from a common ancestor. But do the turbellarians, and therefore the platyhelminths as a whole, form a natural group with a common ancestor? Morphology alone cannot answer this question. Shared primitive characters do not denote phylogenetic relationships; union by negative characters is unconvincing. There is not a single shared secondary ('derived') character that unites the main groups of turbellarians, nor do they share other than primitive characters with the parasitic classes. Catenulids are believed to be near the base of the tree; although simplicity is a very uncertain guide (as the acoelomorphs demonstrate), catenulids are firmly placed by both morphological and molecular evidence as members of the Platyhelminthes. Within the Rhabditophora morphological diversity is greatest in the groups that appear to be the most primitive, which makes it even harder to trace their evolutionary origin. Characters are contradictory: for example, it is primitive in animals for yolk to be inside the egg cell, while in many apparently advanced turbellarians there are separate egg and yolk cells – yet classification according to this character does not accord with classification by the elaboration of the gut. Before these contradictions were resolved by molecular analysis, different investigators emphasised different morphological characters. There has been so much convergent evolution among these flatworms that morphology could not reveal their phylogenetic relationships.

Turbellaria cannot in any case be what is called a monophyletic group, including an ancestor and all of its descendants (see Chapter 2), since the parasitic classes arose from within the group. Since morphological evidence suggested that turbellarians might even be polyphyletic (having no common ancestor), it has been proposed that the class Turbellaria should be abolished. This suggestion was put to one side because the term 'Turbellaria' is so convenient. More recently, molecular evidence has suggested that Turbellaria, including catenulids but not acoelomorphs, may indeed have a common origin (see Chapter 20).

6.6   How are platyhelminths related to other phyla?

Turbellarian morphology is a mixture of simple and complex characters. This led to the belief of some workers that Turbellaria might not be primitive animals but secondarily derived from coelomates by loss of the body cavity, much as leeches, which are annelid worms, have a reduced coelom that has been invaded by tissue. There is, however, little support for the idea that platyhelminths had coelomate ancestors which then lost the coelom, anus and blood system. Developmental evidence is, however, consistent with the placing of platyhelminths among protostome coelomates: cleavage is spiral, much of the future mesoderm is localised in the same early cleavage cells and, although most turbellarians have direct development, the marine polyclads have a larva resembling the annelid trochophore (Figure 9.5). There was, however, insufficient evidence for this placing until it was strongly supported by molecular evidence. What is now clear is that acoelomorphs are very primitive animals at the base of the Bilateria and platyhelminths are not. The traditional assumption that Platyhelminthes represent the earliest triploblastic animals is mistaken.

Chapter 7

Nemertea

Nemertea (also called Nemertinea or Rhynchocoela) are 'proboscis worms', having a proboscis separate from the gut. They are often called ribbon worms since they are typically long and thin (Figure 7.1). They are primitively and predominantly marine, and often beautifully coloured.

Nemertines are not so much a minor phylum as a neglected phylum. Over 1500 species are now known, many having been very recently discovered, and more are being found at a rapid rate now that interest in the phylum has arisen. They are here given

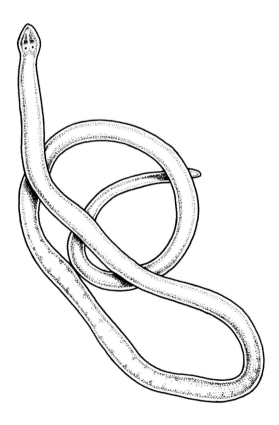

Fig. 7.1 External view of a nemertine.

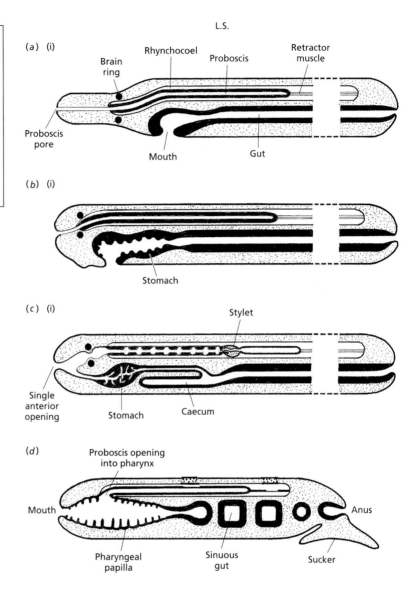

a separate chapter partly because many aspects of their biology, including their controversial evolutionary relationships, are very unusual and worthy of study; partly also because the author's interest in this phylum has been sharpened by her work on terrestrial and freshwater nemertines.

Nemertines share some basic structural features with the Platyhelminthes, from which they were traditionally derived. Although this idea has been abandoned (both phyla are now thought to be related to coelomate protostomes), comparison with platyhelminths is still a good approach to the phylum. Consideration of the characters by which nemertines differ from platyhelminths provides both a definition of the distinctive body plan of nemertines and an introduction to some of the most interesting features of the phylum.

Fig. 7.2 (contd.)

T.S.

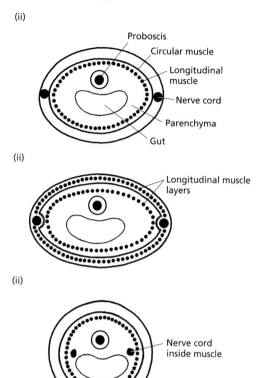

(ii)

(ii)

(ii)

7.1 What are the principal groups of nemertines?

Three orders of nemertines are recognised, within two classes (previously there were four orders) (Figure 7.2):

Anopla. The proboscis has no stylets: the mouth is below or behind the brain.

> **Palaeonemertea**, e.g. *Cephalothrix, Tubulanus*. The nerve cords are peripheral and the worms have many simple features. (This is unlikely to be a natural group.)
>
> **Heteronemertea** e.g. *Lineus, Cerebratulus*. These are the largest and most muscular nemertines; the main muscle layer is the outer longitudinal one.

Enopla. The proboscis bears stylets: the mouth is anterior, opening with the rhynchocoel.

> **Hoplonemertea**, e.g. *Paranemertes, Prostoma, Argonemertes*. The central nervous system is inside the body wall muscle: the

outer muscle layer is circular. *Malacobdella* is a filter-feeding commensal living in the mantle cavities of bivalve molluscs. Once placed in a separate group, 'Bdellonemertea', it is now known to be a highly specialised hoplonemertean.

7.2   How do nemertines resemble platyhelminths?

They have three body layers, ectoderm, mesoderm and endoderm enclosing a gut. There is no general body cavity (i.e. they are acoelomate) and they are unsegmented. They have no rigid cuticle, being bounded by a ciliated epidermis, nor other hard skeleton. They move forward by ciliary movement (if they are sufficiently small) and by muscles squeezing on the semi-fluid unspecialised parenchyma. The anterior end has sense organs such as eyes, and there is a simple brain and nervous system where nerve cell bodies and fibres are barely sorted out. Respiratory exchange occurs all over the body surface. There are protonephridia, which are excretory ducts made from ectodermal intuckings, often with flame cells (see Box 9.1) at the blind endings.

These resemblances do not include any unique common characters. They are presented here not to suggest close relationship, but to define a baseline for the interesting specialisations of nemertines.

7.3   How do nemertines differ from platyhelminths?

Apart from the absence of platyhelminth specialisations (see Chapter 6) the most pronounced differences are:

1. The proboscis
2. The gut has two openings, mouth and anus
3. The presence of a closed blood system
4. Unique cerebral sensory organs
5. Simple gonads, regularly repeated along the body
6. The worms are characteristically very long and thin, and capable of great changes in length.

7.3.1   The proboscis

This lies in the rhynchocoel, a separate fluid-filled cavity dorsal to the gut. Like a coelom it is a space surrounded by mesoderm, but it is not a general body cavity; it contains only the proboscis. Contractions of the rhynchocoel and body wall muscles evert the proboscis and a retractor muscle pulls it back into the worm (Figure 7.3). The proboscis is primarily used in food capture: nemertines are typically active predators. It may also be used, especially in terrestrial nemertines, for rapid escape: it is everted,

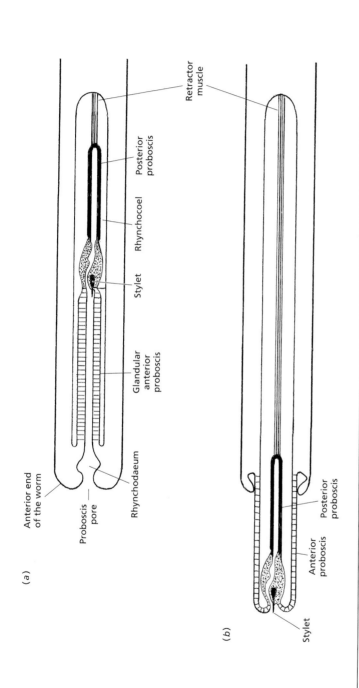

(a)

Anterior end
of the worm

Proboscis
pore

Rhynchodaeum

Glandular
anterior
proboscis

Stylet

Rhynchocoel

Posterior
proboscis

Retractor
muscle

(b)

Stylet

Anterior
proboscis

Posterior
proboscis

Fig. 7.3 Diagrams showing, in dorsal view, the eversion of the hoplonemertean proboscis: (a) proboscis retracted; (b) proboscis everted.

the tip attaches to the ground and the worm pulls itself forward around the proboscis.

The proboscis may or may not be armed with stylets. On this difference, and on the arrangement of the body wall muscle and lateral nerves, rests the main subdivision of the phylum (see above).

7.3.2 The gut

The gut opens at both ends: there is therefore one-way traffic of food and there can be regional differentiation. This not only allows nemertines to become elongated, it also at once demands a blood transport system so that absorbed food can be distributed throughout the worm.

7.3.3 The blood system

The blood system is formed by mesodermal splitting. This method of formation is unique among invertebrates (usually blood is contained in the persisting primary body cavity). The nemertine blood

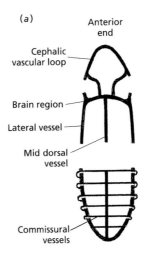

(a)

Anterior end

Cephalic vascular loop

Brain region

Lateral vessel

Mid dorsal vessel

Commissural vessels

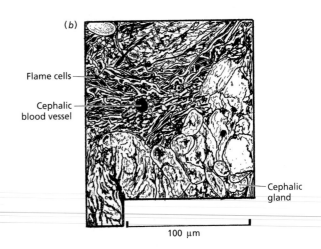

Fig. 7.4 The vascular system of a hoplonemertean: (a) plan of blood vessels, one common version; (b) part of an anterior TS of *Geonemertes pelaensis*, a terrestrial nemertine, showing the accumulation of flame cells round a cephalic blood vessel; (c) a rhynchocoelic villus; (d) TS *Argonemertes dendyi*, a terrestrial nemertine, showing two circular vascular plugs in place of a villus. (b) and (d) drawn from photomicrographs.

(b)

Flame cells

Cephalic blood vessel

Cephalic gland

100 μm

(c)

Fig. 7.4 (contd.)

Proboscis wall

Rhynchocoel

Blood vessel

Villus

(d)

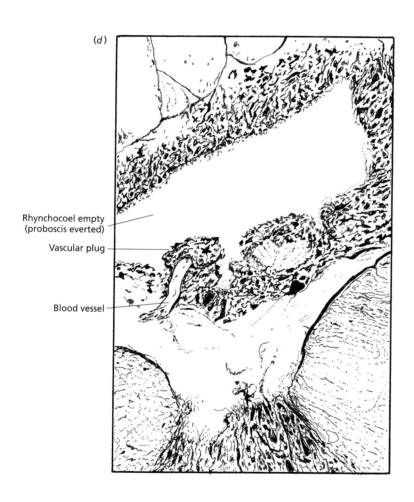

Rhynchocoel empty
(proboscis everted)

Vascular plug

Blood vessel

system appears to be a quite separate product of evolution, made necessary by the second opening of the gut. Typically it consists of a loop at the anterior end with vessels meeting under the brain, and one mid-dorsal and two lateral vessels (often with regular cross-connections) extending posteriorly to join near the anus (Figure 7.4a). There may be an elaboration of vessels beside the foregut, and blood is prominent at any site of water transfer: two examples are that the excretory system is often closely associated with blood vessels (Figures 7.4b and 7.6) and that usually the mid-dorsal vessel penetrates the rhynchocoel wall in the brain region and runs for a short distance bathed in the fluid of the rhynchocoel, the proboscis having no other blood supply (Figure 7.4c,d).

The blood does not seem to be required for oxygen transport and indeed is situated deeply beneath the body wall musculature. Haemoglobin, if present, is usually in the brain and nervous system.

7.3.4 The cerebral organ

This neuroglandular structure is closely associated with the brain (Figure 7.5) and unique to nemertines. It is partly sensory, being concerned with the detection of food, light and the osmotic concentration of the surrounding water. It also regulates the secretion into the blood of acid mucopolysaccharides from which mucus is made. We need more experimental work concerning the functions of this unusual organ.

7.3.5 The gonads

Nemertine gonads are simple and very numerous: they are regularly repeated between lobes of the gut along the length of the worm, giving these worms an almost segmented appearance. Sexual reproduction predominates, but some species have great regenerative powers that can lead to asexual reproduction. Most nemertines have separate sexes but some species (including all those known from fresh water) are hermaphroditic. Most commonly eggs and sperm

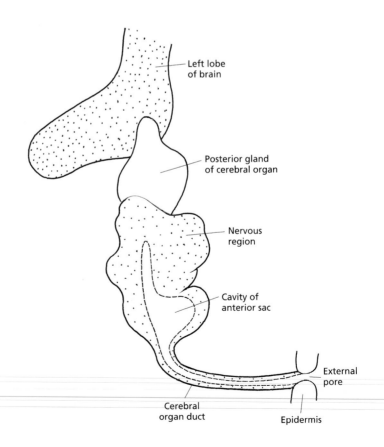

Fig. 7.5 *Argonemertes dendyi:* dorsal view of the left cerebral organ.

from different animals are shed from the many gonads and external fertilisation occurs in the water. The embryo may develop directly within the egg, or the egg may hatch into a planktonic larva.

7.3.6 Length and extensibility

Nemertines are often very long and thin, and very extensible. Some (e.g. terrestrial and interstitial species) are only a few millimetres long, but some in the sea are very long indeed: for example *Lineus longissimus* can reach 30 metres, while remaining only a few millimetres in diameter. What is the advantage of so great an extension of one animal? There will be a usefully large number of gonads, but why attach them all to one feeding head? The answer probably lies in the nature of the epidermis. It is deeply folded with microvilli and many mitochondria, and in some species experiments with radioactively labelled glucose and amino acids have shown that the epidermis can absorb dissolved organic food as a supplement to normal feeding. Extension in length is therefore a structural economy. A long worm might be more vulnerable to predators, but the epidermis secretes deterrent toxins. A crab taking a nemertine may hold it in one claw and wipe it clean with the other before eating it: not all predators are so wary.

Nemertines are not only often long, they are capable of unusual changes in length: by a factor of 10 in *Lineus*. The lateral nerve includes a longitudinal muscle fibre that assists these changes. Proboscis eversion can involve a 50-fold change in length of the proboscis retractor muscle: the proboscis can be shot out (by fluid pressure caused by muscle contraction) and pulled back (by the retractor muscle) very fast. This combination of great change in length with rapidity of contraction is a most unusual property for any muscle. There has been no recent experimental work on nemertine muscle.

Could there be a connection between contractility and the amount of muscle in the body? 'Hoop stress' to burst a cylinder sideways can be shown to be twice the longitudinal bursting stress (frying sausages burst along their length, they do not break down into little rings). A worm should therefore have twice as much circular as longitudinal muscle in cross section. Nemertines do not; nor (unlike earthworms) do they use the wall of a burrow for support. Is there some correlation here with very variable contraction?

Other nemertines such as *Cerebratulus* are much thicker and vary little in length. They can swim strongly by vertical undulations passing along the body, which is stiffened by dorsoventral muscle contraction and so does not change in length: leeches do the same thing. Remarkably, if a *Cerebratulus* is suspended half in air and half in water, the part in water will swim while the part in air carries out earthworm-style peristalsis. Clearly there can be very little whole-body coordination of muscle contraction: nemertine muscle and its control should repay further study.

7.4 What diversity exists among nemertines?

Having seen that these worms have many unexpected features, let us look at what natural selection has produced within the phylum.

7.4.1 Morphological diversity

At first sight this is a rather uniform phylum, but there is great and seemingly arbitrary variation on a small scale, sometimes unexpectedly combined in different animals: for example, a curious 'wickerwork' construction of the rhynchocoelic muscle has evolved several times, and one species in the otherwise unarmed class Heteronemertea has many small stylets on the proboscis. Occasional species have characters quite unexpected for the phylum: one has ganglia along a dorsal nerve, some have striated muscle and one tiny species living between sand grains shows external annulations, rather like an earthworm. There has clearly been a great deal of convergent evolution. Even the simple classification given above is controversial: emphasis on different characters produces different hypotheses about evolutionary relationships. On any scheme, some elaborate and apparently stable characters must have evolved more than once.

7.4.2 Feeding methods

These at once illustrate the theme of much small-scale diversity superimposed on apparent uniformity. Nemertines characteristically feed as carnivores: they may be ambush predators or may actively hunt their prey, which commonly consists of crustaceans or polychaete (marine annelid) worms. The proboscis is everted rapidly and accurately and wraps around the prey, which in unarmed species is swallowed whole. Armed nemertines paralyse or kill their prey by piercing it with stylets and secreting toxins. Yet nemertines do not always rely on the proboscis: some absorb organic food in solution all over the surface, for example *Lineus longissimus* (see above) and the larval stages of *Carcinonemertes*, an egg predator living on crabs; its adult stages do use the proboscis (they will only evert it on a surface curved like an egg). *Malacobdella* is a commensal (i.e. it shares the food of another animal with which it lives). Here the proboscis is less important and neither mucus nor toxin is secreted. The pharynx expands and the papillae projecting from its inner surface interlock, making a net that entraps food particles. Compressed, the water then passes backward, propelled by cilia at the base of the papillae. This is feeding like a flamingo or a baleen whale.

7.4.3 Ecology

Ecological diversity is considerable. Nemertines are found in all marine habitats from the deep seas to the top of the shore: they are common in both tropical and polar seas. They may live as commensals with tunicates or anemones or as egg predators under the carapaces of crabs. About a dozen species have colonised land, living in damp cool places, for example under rotting logs. Terrestrial

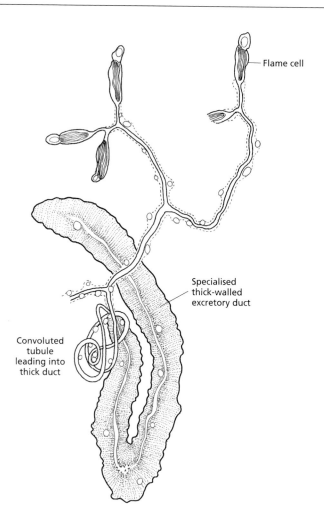

Flame cell

Fig. 7.6 *Argonemertes dendyi*: one protonephridium.

Specialised
thick-walled
excretory duct

Convoluted
tubule
leading into
thick duct

adaptations have arisen several times separately, in species from different hoplonemertean families, showing strong convergence: for example all have great development of the excretory system. Figure 7.6 shows one of the many thousand protonephridia in a terrestrial nemertine some 10 mm long. This system removes excess water entering when the worms are flooded, and extensive mucus production which protects them against desiccation. Rather more species, of both heteronemerteans and hoplonemerteans, are found in fresh water, again showing great convergence in a number of characters. There appear to have been two routes for colonisation of fresh water, one by way of estuaries and the other overland.

7.5 How do nemertines develop?

Early development in most species is by spiral cleavage. The blastopore forms the mouth before the anus, as in protostomes such as platyhelminths, annelids and molluscs (Box 5.2). Future mesoderm is determined early, and derived from the same early cells as in the coelomate protostomes (see Chapter 19).

Most nemertines have direct development. A number of heteronemerteans, however, have a larval stage, the pilidium, which (like the larva of marine platyhelminths) has many resemblances to the annelid trochophore. Direct development has been assumed to be primitive for the phylum; careful recent work, however, has detected some form of 'hidden larva' in other groups, including hoplonemerteans and even palaeonemerteans, where *Carinoma* shows evidence of a 'prototroch' (the first band of cilia in a trochophore). This is another similarity to annelids in the early development of nemertines.

7.6 How are nemertines related to other phyla?

No certain fossil evidence exists, and evidence from morphology and development has generated much controversy. The traditional derivation from a platyhelminth-like form is no longer accepted. Molecular evidence strongly supports the evidence from early development that places nemertines close to coelomate protostomes rather than relating them to early acoelomate bilaterians. Nemertines have no general body cavity, but from morphology the rhynchocoel can be described as a restricted coelom, serving only to house the proboscis and generate hydrostatic pressure for its eversion. It is not a body cavity, however, and its development differs from that of a coelom. Again, the blood system is formed by splitting of mesoderm and is lined by a mesodermal epithelium, yet the cavity enclosed (inside the blood vessel) is extremely small. Are these cavities special nemertine features converging on coeloms or are they remnants of ancestral body cavities? Molecular evidence can help to establish relationships, but it cannot prove the direction of evolution. Proof is lacking, but regression followed by radiation into a phylum of more than 1500 species seems very improbable: there are no suggestions about why it should be favoured by natural selection, in a phylum consisting mostly of active predators where a coelomic body cavity would promote rapid locomotion, and no evidence that it has occurred. Partial 'coelomic' cavities are known to have arisen many times in the evolutionary history of other phyla, and convergence remains the most probable explanation of the nature of these nemertine cavities.

There is no convincing evidence that any other phylum has arisen from early nemertines. It has been claimed that they gave rise to other phyla, even to chordates and hence to vertebrates, but these ideas have never been accepted by most morphologists and are contrary to the molecular evidence, which totally supports the division between protostomes and deuterostomes, with nemertines as protostomes. Convergent evolution of certain characters can account for any resemblances.

Box 7.1 | Sea, fresh water and land

Sea

Animal life began in the sea. The marine habitat is favourable, being a large body of water fairly constant in quantity, movement and salt concentration. It is buoyant, relatively rich in potential food and it enables animals to disperse. Many marine invertebrates shed their eggs and sperm directly into the sea and fertilisation is external. The eggs contain little yolk and hatch early, usually into larvae feeding in the plankton (surface waters). Above all, the body fluids of marine invertebrates are similar in salt concentration to that of the sea in which they live, and there is no osmotic problem.

Units and definitions

The **osmotic pressure** of a solute is usually measured by the depression of the freezing point (Δ) caused by the solutes. Pure water freezes at 0.00 °C; sea water (with 35 parts of salt per 1000) at −1.86 °C.

Osmoles measure the osmotic effect of a solute whose molecular weight in grams is dissolved in a litre of sea water. For sea water the value is 1 osmole (= 1000 milliosmoles or mosm) and this provides a useful standard for comparison.

Most marine invertebrates are **osmoconformers**, and are fully permeable to water, i.e. the body fluid concentration is close to one osmole.

Osmotic regulation by contrast is the maintenance of the internal body fluids at a concentration different from that of the environment.

Euryhaline animals are those which can tolerate a wide range of salinities, usually by a combination of osmoconforming and osmoregulating processes.

Stenohaline animals are those (like most marine invertebrates incapable of osmoregulation) that are restricted to a narrow salinity range.

Ionic regulation, selection of particular ions at concentrations different from those in the sea, does however occur even in the simplest marine animals. A jellyfish floats easily because it contains relative to the surrounding sea fewer heavy ions (magnesium, sulphate) and more lighter ions (sodium, chloride). A marine lobster (*Homarus* sp.) maintains concentrations of magnesium ions at 14% and sodium at 111% sea water, but it is incapable of osmoregulation.

Fresh water

Colonisation of fresh water from the sea raises problems. The medium is much more variable in many ways: it may over-heat or freeze, it may flow very fast, it may dry up, and above all it is much more dilute (5–10 mosm) than internal fluids can be. Water will therefore tend to enter by osmosis, and salts (ions) to leave by diffusion. If these movements were unchecked, animals would swell up and burst or all the salts would leak away. Small animals, with relatively more surface area in proportion to their volume, have a worse problem than larger ones.

Long-term adaptations to life in fresh water include:

1. **Reduction of the permeability of the outer surface**. Some permeability must remain, if only in the gills, to allow intake of oxygen for respiration and removal of carbon dioxide.

2. **Reduction of the internal concentration** to reduce the difference between inside and outside. The body fluids of freshwater animals are almost always more dilute than those of their marine relatives (perhaps 100–300 mosm) but they cannot be as dilute as the medium.

3. **Tolerance of fluctuations** in the salinity of the body fluids is characteristic of many estuarine osmoconformers, for example annelids, molluscs and crustaceans. Estuarine animals need to be euryhaline. One common adaptation is to retain amino acids within the cells: when the body fluids are diluted by osmotic influx of water, the cells release the amino acids, becoming temporarily as dilute as the blood and therefore not subject to entry of water.

The above adaptations are insufficient by themselves to support life in fresh water: osmoregulation, an energy-consuming process, is required. Water is never transported actively; ions are moved, and water follows by osmosis:

1. **Active uptake of ions at the gills or over the whole body surface** occurs in all freshwater animals. Usually sodium or chloride ions, or both, are taken in, compensating for the outward diffusion of salts. Entering water is removed by the excretory organ (see diagram).

Diagram showing salt and ion movement in freshwater animals. Note that active ion reabsorption in the excretory organ occurs only in some of the freshwater animals.

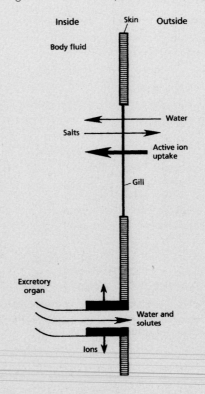

2. **Active reabsorption of ions in the excretory organ**, with the production of urine more dilute than the blood (hypotonic), may reinforce ion uptake at the surface. This occurs in many but by no means all freshwater animals.

Reproductive adaptations

Sperm do not survive well in fresh water and external fertilisation is rare. More freshwater animals are hermaphrodites. The eggs typically contain much yolk and most develop directly: a fragile larva, small with a relatively large surface area, would be very vulnerable to osmosis/salt loss and liable to be swept downstream.

Fresh water was colonised only in part by the aquatic route up estuaries. Many freshwater animals (nearly all of the molluscs and arthropods) evolved from terrestrial forms secondarily returning to water.

Land

Many so-called terrestrial invertebrates live in soil (e.g. earthworms, planarians, nemertines, nematodes) or other moist places, and remain physiologically aquatic.

Problems for animals colonising land include:

1. Avoidance of desiccation
2. Tolerance or avoidance of temperature fluctuations
3. Support and locomotion of the body in air.

Not surprisingly, most truly terrestrial invertebrates are arthropods or molluscs.

When surrounded by dry air rather than water animals must:

1. Respire using lungs rather than gills (see Box 9.2)
2. Excrete with minimal water loss (see Box 9.1)
3. Perceive different environmental stimuli, often with different sense organs.

Routes by which land was colonised

Many small animals (about 2 mm long) came by the interstitial route, evolving from ancestors living in the spaces between grains of sand or other sediments. Most terrestrial animals, however, came to land either by the aquatic route from fresh water (e.g. vertebrates) or by the overland route from the top of the seashore (e.g. most terrestrial arthropods and many molluscs). The routes may be identified by differences between present-day animals:

1. Animals once adapted to fresh water have body fluids more dilute than the sea and this difference will be maintained in their further evolution (e.g. the body fluids of slugs and snails range from 97 to 231 mosm).
2. Terrestrial animals whose ancestors took the overland route are likely to have body fluids nearer to marine concentrations (e.g. 700 mosm in the woodlouse *Porcellio*). Animals evolving by way of the top of the seashore are also likely to tolerate some evaporative water loss.

Chapter 8

Nematoda

Nematodes, or 'roundworms', are totally different from platyhelminths and nemertines. A nematode is characterised by an extremely tough thick cuticle round the outside and a very high hydrostatic pressure within. They all look very similar (Figure 8.1a) and given the pressure it is hard to see how any other shape could be maintained, yet there are perhaps a million species. The phylum is unusually ubiquitous; nematodes are free-living in marine, freshwater and land habitats and parasitic in animals and plants. They are clearly of great economic importance as they exist in extraordinary numbers and play a significant role in the total matter and energy cycle of the biosphere. Buchsbaum (*Animals Without Backbones*, 1938) wrote, 'If all the matter in the universe except the nematodes were swept away, our world would still be dimly recognisable, and if, as disembodied spirits, we could then investigate it, we should find its mountains, hills, vales, rivers, lakes and oceans. The location of towns would be decipherable ... Trees would stand in ghostly rows', etc. We are told that four and a half million individual nematodes were found in one square metre of Dutch marine mud, and 90 000 in one rotten apple.

What makes these worms so remarkably widespread and numerous? The cuticle is clearly the answer, and it is argued below that the combination of tough cuticle and high internal pressure dictates most of their (often unique) characteristics. Their development is also discussed: like their morphology it shows unusual uniformity in that a nematode has a fixed number of large cells of predetermined fate.

Parasitic nematodes have been very harmful to humans in causing disease and in damaging agriculture: there are many accounts of the particular importance of nematodes as ubiquitous, numerous and successful parasites with remarkably varied life cycles. The present introduction to the phylum concentrates instead on the unique characteristics of nematodes and the opportunities which these have provided for research on the control of development by genes. Some of the results of this research can be applied to humans and (as will be indicated) have provided us with important new medical tools.

8.1 What are the distinctive characters of nematodes?

Nematodes are slender worms, circular in cross-section and ranging in length from 200 μm to 40 cm (in *Ascaris*, a large intestinal human parasite) or even 9 metres in a parasite of whales. The worms are triploblastic, unsegmented and enclosed in a tough but flexible cuticle, under which is a layer of muscle, longitudinal only. There is no blood or other circulatory system. The body cavity has no inner lining of mesoderm and is not a coelom; it is derived directly from the blastocoel (see Box 5.2). It contains the many tubes of the reproductive system and also the gut, which opens at each end of the worm. The general structure (Figure 8.1b,c) is a series of tubes within a tube.

8.2 How are these characters related to the cuticle and fluid pressure?

8.2.1 The cuticle

This organ is a many-layered structure toughened by cross-linked protein chains and incorporating a collagen lattice which enables movement (see Chapter 5). Figure 8.1d indicates the structure in *Ascaris*, with nine layers in three zones. The cuticle is laid down by an underlying 'hypodermis' (the equivalent of the outer epidermis of other worms) which is 'syncitial'. The cuticle is extremely resistant both mechanically and chemically: it enables nematodes to survive in deserts and polar regions and to withstand the digestive enzymes of host animals. It is undoubtedly fundamental to the success of nematodes as parasites and it has been very hard to develop pesticides effective against them.

8.2.2 The body cavity

The body cavity contains fluid at pressure that may be an order of magnitude higher than that of an earthworm, which operates at 3–30 cm water (0.28–2.8 kN per square metre). The figures for *Ascaris* are 70–400 cm water (6.6–37.6 kN per square metre).

8.2.3 The locomotor system

The locomotor system depends entirely upon muscle working against the cuticle and the internal pressure. There are no cilia: they could beat neither on the outside nor on the inside. Movement is achieved by contraction of four blocks of longitudinal muscle, which

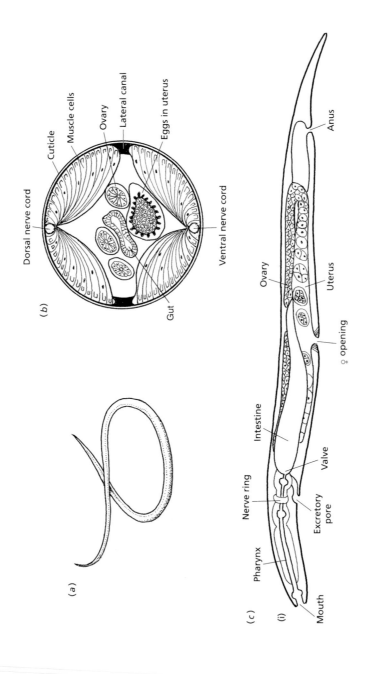

Fig. 8.1 The structure of nematodes: (a) external view; (b) transverse section based on *Ascaris*; (c) longitudinal section of (i) female, (ii) male, based on *Rhabditis*; (d) nematode cuticle.

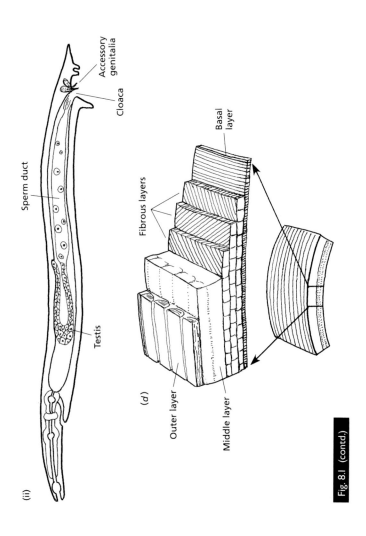

(ii)

Accessory
genitalia

Cloaca

Sperm duct

Testis

(d)

Fibrous layers

Basal
layer

Outer layer

Middle layer

Fig. 8.1 (contd.)

(most unusually) send processes to the nerves rather than receiving nerve endings. The muscle is further unusual in having contractile and non-contractile regions (the contractile parts are obliquely striated with a pattern of thick myosin and thin actin filaments, as shown in Box 5.1). The cells are very large: a single muscle cell may be 10 mm in length. Circular muscles are absent: they could not contract under the cuticle and are in any case redundant, because longitudinal contraction on one side of the worm is opposed by pressure making the cuticle bulge on the other side, guided by the collagen lattice, without overall change in shape. Since there is no circular muscle, the lattice angle must at all times be greater than 55° (see Chapter 5). Each half of a piece of the worm alternately contracts and relaxes, throwing the animal into lateral waves. These waves are propagated backward along the worm, making it move forward with the undulating progression that is so characteristic of nematodes (Figure 5.1c). Locomotion is only possible when there is soil or tissue for the body to press against: a freely floating nematode will thrash about ineffectively.

Coordination in such a system is achieved largely by hydraulics. There is a minimal nervous system, with longitudinal nerve cords and an anterior nerve ring in place of a brain. Sensory structures may be present at either end as small projecting tactile papillae or chemosensory pits.

8.2.4 Other systems

Other body systems are similarly governed by the cuticle and high internal pressure. Feeding is by a mouth pump, sometimes assisted by stylets, drawing in semi-liquid food. A strong muscle closes the anus to keep the food inside. Respiration is often largely anaerobic, based on partial breakdown of stored glycogen (animal starch). In a classic early experiment, nematodes were buried in anoxic mud from the Clyde estuary in Scotland and found still to be active after 35 days. Where the environment can provide oxygen, this diffuses in over the body wall and combines with the haemoglobin (which has a particularly high affinity for oxygen).

Excretion has been attributed to two large 'renette' cells that may be elongated into tubes running the length of the body, with a cross-connection and a pore. However, experiments showed that urea is voided not by the tube's pore but by the anus. The ultrastructure of the tube resembles that of the contractile vacuole of Amoeba, suggesting an osmoregulatory function (see Box 7.1). Note that while function cannot safely be inferred from large-scale structure, at the smallest level structure and function are so closely associated that such inference becomes respectable.

8.2.5 Reproduction and life history

Reproduction is always sexual: regeneration and budding cannot occur. Most of the space inside a nematode is filled by the reproductive system. The sexes are usually separate, and fertilisation is internal. Again the high internal pressure makes problems: the oviducts have to be kept open with spicules supplied by the male penis. The sperm are not flagellated in nematodes (probably flagella could not beat against the pressure) nor is 'amoeboid' a precise description of their movement. They 'crawl', aided by a specially secreted protein. The numerous 'eggs' laid are in fact zygotes or early embryos: they hatch as sexually immature miniature adults.

The growth of the young nematode is punctuated by moults, although the cuticle does allow some increase in size. Consequently the different stages are described as larvae, although their similarity to the adult makes 'juvenile' a better term. In parasites the third of these four stages is characteristically a wandering stage, when a new host or part of a host is colonised. A large volume of reproduction is most important for a parasite (see Box 6.1) and an elaborate life history may facilitate transfer between hosts.

8.3   How is the phylum subdivided?

The unusual degree of morphological uniformity makes classification extremely difficult. There is no help from fossils, which are few and recent. Primitively free-living and probably marine, nematodes (unlike platyhelminths) seem to have evolved parasitism many times independently, with far-reaching convergence making it even harder to trace their evolutionary relationships. Many families have been named, and recently molecular methods have made much progress, but identification of nematodes remains a task for experts.

8.4   Why are nematodes useful for developmental studies?

Part of the zoologist's delight in studying animal diversity is to pick the animal for a 'model system', that is, to yield answers to puzzling questions or information when complexity seems formidable. Nematodes were the inspired choice of Sydney Brenner in Cambridge in the early 1960s, for studying animal development. His initial work, published in 1974, was extended by Horvitz and many others, in a team later led by Sulston. Brenner, Horvitz and Sulston shared the Nobel Prize.

The early development of nematodes has long been known to be very unusual. In the late nineteenth century Theodor Boveri studied a species of *Ascaris* with two chromosomes and observed 'chromosome diminution', the loss of parts of them in early cell divisions. We now know that at the 16-cell stage a full set of genes is preserved in only two cells, one of which will give rise to the germ cells. In the future somatic cells, the protein composition of the ribosomes is changed by this loss of material from the chromosomes – a unique method of gene regulation. In Boveri's time this procedure was not seen as particularly odd, and emphasis fell on the unique 'nematode T', a temporary configuration of the first four cells (see Figure 19.3).

Modern study has focused on the remarkably small number of cells in an adult nematode and the fixity of their developmental origin. There is an invariant sequence of cell divisions, each determining the nature and future position of the daughter cells. Early in development cell divisions cease, so that for a particular species there is a fixed number of cells both in a given tissue and in the whole nematode (about 1000 somatic cells on average). Further enlargement of the worm is due solely to increase in size of its cells, which may become remarkably large in a large adult. Even more remarkably, the fate of each cell depends upon its lineage; most developing animals override information from cell lineage with regulating messages between neighbouring cells. It is these peculiarities that make nematodes unusually accessible to experimental study of development and its genetic control.

As so often in biology, the above generalisations have to be qualified because natural selection is at work. The picture of 'determinate' development with lineage deciding the fate of cells is correct and important, and carried to a greater extreme in nematodes than in any other animals known, but there are exceptions. Some tissues (the hypodermis and the gut) of some species do have variable cell numbers, and careful studies have revealed regulatory action between cells at various stages in development, even between the first two cells produced when the zygote divides. Regulative cell interaction was first recognised in the formation of the vulva, a non-essential structure because in its absence the hermaphrodite can fertilise itself. Two adjacent cells in the female gonad interact so that one of them becomes the uterine precursor and the other the 'anchor cell'; this cell first initiates the cell lineage that makes a vulva and then induces the ventral hypodermis to make it an external opening. Although many examples of regulation are now known, determinate development predominates, and the fate of a cell can usually be predicted from its lineage. Since nematodes have a small and finite number of cells which become unusually large, a whole embryonic region may be represented by a single cell rather than a group of cells. The developmental potential of that cell is then severely limited.

8.5 Why has *Caenorhabditis elegans* been studied so thoroughly?

Caenorhabditis elegans (Figure 8.2) is a free-living nematode in the soil, 1 mm in length and transparent. It can be cultured in the laboratory

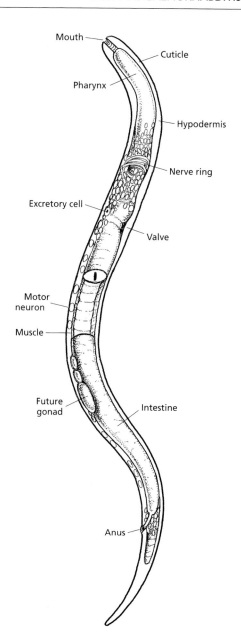

Mouth
Cuticle
Pharynx
Hypodermis
Nerve ring
Excretory cell
Valve
Motor neuron
Muscle
Future gonad
Intestine
Anus

Fig. 8.2 *Caenorhabditis elegans: young specimen, with internal structure seen by transparency.*

and at 20 °C it completes its life cycle in 3.5 days, producing adults that are either hermaphrodites or males. Brenner picked out *C. elegans* as an organism where it is possible to identify every cell as it develops and to trace its lineage; since the worms are transparent, observation of living cells can give continuity and certainty (and more recently observation has been combined with videotape recording). The complete cell lineage is now known: the earliest steps in the sequence are shown in Figure 8.3.

C. elegans was the first invertebrate to be sequenced: the entire genetic sequence of more than 19 000 genes is now fully known,

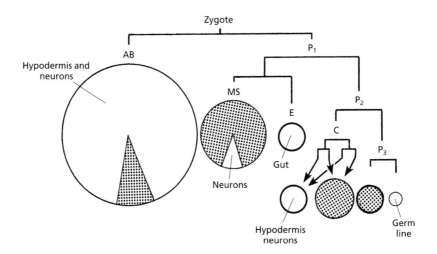

Fig. 8.3 *Caenorhabditis elegans*: the first divisions of the zygote, indicating how the future fate of the daughter cells is progressively determined at each division: P_{1-3} represents the germ line to future gonads and gametes. Muscle develops from all four shaded areas. Note multiple origin of neurons also.

and more recently a second nematode has been sequenced for comparison. The genetic control of the developmental fate of each cell is largely understood, in terms of a hierarchy of cell decisions governed by particular genes, making regulatory proteins that determine the effects of other genes.

Some interesting findings are:

1. At hatching the hermaphrodite has 550 non-gonadal nuclei, increasing to 810 at maturity. The mature male has 959. Cell numbers for different tissues are: nerve 302, muscle 122 and intestine only 34. An animal with very few cells may need to be able to use them for more than one function, and this is characteristic of nematode cells. The cells have also remained able to evolve great differences in form and function. These features must have contributed to the great success of nematodes.

2. Cells of a kind do not always arise from the same germ layer (i.e. ectoderm, mesoderm, endoderm). Identical muscle cells arise from quite distant positions; nerve cells are formed at three separate stages in the lineage (Figure 8.3).

3. Cell death is programmed to occur at certain stages in the lineage.

Genetic analysis is based on developmental mutants. A crucial advantage of *C. elegans* is that genetics can be combined with analysis of the animal at single-cell resolution. The ability of the hermaphrodite to be either self- or cross-fertilised is a further useful feature. Cell lineage defects in mutants can be precisely described, and alterations can be understood in terms of altered developmental decisions in individual cells. Mutations may result in discrete transformations of cell fate at any point in a lineage, or may switch development between lineages. Powerful probing of *C. elegans* revealed ten times more mutations than had been predicted, the first clear demonstration that mutations are not as rare as theory had dictated. This finding proved to have general application and interesting causes, as will be explained in Chapter 20. Genomes as a whole, in contrast to individual genes, have greatly changed in the course of evolution.

The extent to which genes, and therefore fundamental biological processes, have been conserved throughout evolution is the most surprising and important discovery to emerge from comparative work on nematodes, fruit flies and mammals. Even comparison of the genome of *C. elegans* with that of yeast (a unicellular fungus) reveals a considerable degree of resemblance; the differences may help to give us a genetic definition of a multicellular animal. At the same time, about 400 genes are found that are specific to nematodes, for example those concerned with biochemically unique structures such as the cuticle. Such genes could be useful to control nematode pests without damaging other animals. *C. elegans* has equivalents of about half of known human genes, some of which offer medical opportunities: the principle is that when a gene is known to cause an illness, that gene can be inserted into the worm to find exactly what goes wrong. To give a few examples of the potential use of opportunities offered by work on the nematode:

1. Genes controlling programmed cell death in *C. elegans* may become important in treating cancer (where cells fail to die).
2. One gene is found to alter the rates of metabolism and of ageing throughout the nematode, and this gene has a human counterpart that makes a receptor for insulin.
3. The recently discovered form of RNA called RNAi has been made accessible by work on *C. elegans*, where it is found to control the patterning of neurones. RNAi consists of small pieces of double-stranded RNA that can inhibit the expression of the genes from which they are derived, by destroying their mRNA. An insertion that can 'shut genes off' is clearly a wonderful tool for investigating gene function: RNAi has been used to silence nematode genes that correspond to mammalian disease-causing genes. The way is open to medical suppression of such genes without damage to the living human body in which they occur.

8.6   How are nematodes related to other animals?

Morphological evidence has failed to answer this question, and molecular comparisons have been hampered by the unusually rapid rate of nucleotide substitutions in nematode ribosomal DNA. However, emerging evidence rather surprisingly places the phylum with arthropods and other animals with moulted cuticles (see Chapter 20).

8.7 Conclusion

Nematodes are exceptionally uniform and widespread, combining uniformity of structure with variety of habitat. It has been said that there appears to be only one nematode, the model coming in various sizes and with a great variety of life histories. Their success and their morphological uniformity can both be attributed to the unique structure and properties of the cuticle. Their developmental uniformity, with a fixed hierarchy of cell decisions, has been particularly useful to biological and medical research. Our uniquely full understanding of the genetic regulation of development in nematodes is increasingly applied to investigation of other animals, including ourselves.

Chapter 9

Annelida

Most of the 15 000 known species of annelid worms live in the sea, crawling under rocks on the shore and sea bottom, freely swimming or seeking protection from predators in burrows or tubes. Freshwater and terrestrial annelid radiation has been small except for the earthworms, which are outstandingly successful, and the leeches, which are widespread specialised suctorial predators. Annelids vary in length from 1 mm between sand grains to 3 metres in some Australasian earthworms.

Annelid worms are coelomates with metameric segmentation. The annelid coelom is a large fluid-filled body cavity surrounded by mesoderm, providing an efficient hydrostatic skeleton. Metamerism is the serial repetition of similar parts along the length of an animal, manifested primarily in the separation of the mesoderm into segmental blocks of muscle. In annelids there are usually internal partitions (septa) between the segments. Both the coelom and metamerism improve the effectiveness of muscle contractions, so that active locomotion can be much faster than in acoelomate animals. At the same time these advances demand greater complexity: when an extensive coelom separates the inner and outer tissues, a transport system is required and more elaborate respiratory and excretory organs may need to develop. This complexity then allows more structural differentiation and an increase in size.

The following account first defines and subdivides the phylum, then describes locomotion to show the advantages of coelomate and metameric organisation. The blood, respiratory and excretory systems are introduced, with more background given in Boxes 9.1, 9.2 and 9.3, to illustrate the increased complexity required by a coelomate animal. Reference to feeding and reproduction indicates the diversity within the phylum, and finally the uncertainties about annelid origins and interrelationships are indicated.

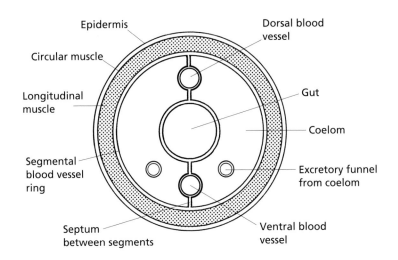

9.1 What is an annelid?

An annelid is a protostome worm with three cell layers, a gut with a mouth and an anus and a body wall with both circular and longitudinal muscle. The coelomic body cavity is formed by the splitting of embryonic mesoderm. The outer epidermis is covered by a thin cuticle, typically bearing chitinous bristles ('chaetae' or 'setae'). Metameric segmentation, always shown in the muscular and nervous systems, is characteristically also evident externally. The nervous system has a supra-oesophageal ganglion (i.e. group of the main bodies of nerve cells), which is called a brain although it may be little more than a sensory relay, and a ventral nerve cord bearing segmental ganglia giving off segmental nerves. There is a closed blood system with the blood moving forward in the dorsal longitudinal vessel. Segmental ducts between the coelom and the outside are used for excretion and reproduction. Figure 9.1 shows the basic annelid structure.

Box 9.1 | Excretion

Narrowly defined, excretion is the removal of the waste products of metabolic activity. Since excretory organs are often involved in osmoregulation (in fresh water) or water conservation (on land), the definition may be extended to include more general maintenance of a fairly constant internal environment, by adjustment of the water and ion content of the body fluids. See Box 7.1, relevant throughout to excretion.

Excretory products

Other than water and ions, excretory products include:

Carbon dioxide, CO_2, a product of respiration, in all animals.

Ammonia, NH_3, ion NH_4^+, the primary product of deamination of amino acids. It is toxic and requires much water for its removal.

Urea, $CO(NH_2)_2$, less toxic and requiring less water loss; it is a very common end point of nitrogenous metabolism, produced by condensing one molecule of CO_2 with two molecules of ammonia.

Uric acid, $C_5O_3N_4H_4$, the end point of purine metabolism; it is a more complex molecule, more expensive to make. It can however be excreted semi-solid with very little water loss, or stored harmlessly within animals or eggs. It is commonly produced by terrestrial animals, for example insects and snails (spiders use another purine, guanine).

Most animals excrete a mixture of these substances (and others), but one nitrogenous end point usually predominates. Ammonia, for example, is 80% of the total in the annelid *Aphrodite*, 60% in the crayfish *Astacus*, 67% in the cuttlefish *Sepia* and 39% in the starfish *Asterias* (where most of the nitrogenous waste is in the form of amino acids). In the land pulmonate snail *Helix* uric acid constitutes 70% of nitrogenous waste, while in the lower shore periwinkle *Littorina littorea* it is only 0.8–1.2%. Woodlice are exceptional among terrestrial animals in being able to puff out gaseous ammonia.

Excretory organs

Sponges, Cnidaria and echinoderms have no excretory organs. Many other marine invertebrates may give out ammonia as well as carbon dioxide over the body surface or the gills, as well as through excretory organs. Most excretory organs are nephridia, ciliated excretory tubules leading from the internal tissue or body cavity to the outside world. Typically, fluid enters the tubule by ultrafiltration and its composition is modified as it passes along the tubule; it opens to the outside world directly by a pore or indirectly by way of a collecting duct. Nephridia occur in most animals other than nematodes and terrestrial arthropods. There are two kinds:

A **protonephridium** has a blind inner end bearing a flagellum or a tuft of cilia, beating to draw water and solutes into the tubule. Protonephridia take the form of flame cells (see diagram, and Figure 7.6) or similar structures; they occur in platyhelminths, nemertines, priapulids, gastrotrichs and many larval and a few adult polychaetes.

A **metanephridium** is open at both ends. The inner end has a ciliated funnel which filters fluid from the body cavity into the tubule (see diagram). Metanephridia are the basis of the excretory organs of most adult invertebrates. Although not confined to segmented animals they are often called 'segmental organs'; they may serve as gonadal as well as excretory ducts.

Originally nephridia were defined as ectodermal intuckings, as opposed to mesodermal 'coelomoducts' but the originating tissue may come from either layer. Protonephridia are thought to have evolved once only, while meta-nephridia have evolved more than once by different developmental pathways.

How excretory organs work

Two processes are involved:

Ultrafiltration. Pressure forces water and small dissolved molecules through a semi-permeable membrane that holds back large molecules such as proteins.

Active transport of ions (never of water) is a necessary second step: the filtrate in the excretory tubule is modified as it passes along, selected substances being added or reabsorbed. While normally it is superimposed on the filtration process, active transport may also initiate excretion if, as in the insect haemocoel, blood pressure is lacking.

Transverse section (TS) of (a) acoelomate worm with two proto-nephridia (the left one is shown as a flame cell and both are dispropor-tionately large); (b) flame cell; (c) coelomate worm with metane-phridium; (d) duct of (c), enlarged.

(a) Protonephridia

(b) Nucleus
Ultrafiltration site
Ciliary 'flame'
Fluid modified near pore
Tubule
Pore

(c) Coelom
Metanephridium

(d) Ultrafiltration
Ions secreted or reabsorbed
Pore

9.2 What annelids are there?

The three main classes are:

Polychaeta. This large class of marine annelids shows great diversity. Characteristically annelids bear 'parapodia', paired lobes on the sides of most segments forming paddle-like appendages with bundles of chaetae (Figure 9.2a,b).

Oligochaeta. Mainly freshwater or terrestrial annelids, without parapodia and with a few unjointed chaetae (Figure 9.2c,d). The coelom is large and spacious. Feeding is suctorial; there are no jaws. Oligochaetes are characterised by a saddle-like 'clitellum', an epidermal thickening that secretes a cocoon.

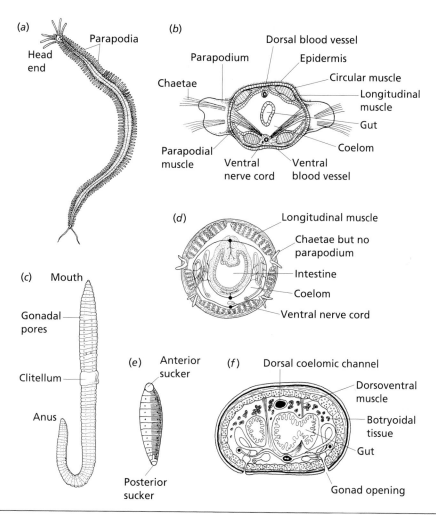

Fig. 9.2 (a) A nereid polychaete; (b) transverse section (TS) of a nereid; (c) the earthworm *Lumbricus terrestris*, an oligochaete; (d) TS *Lumbricus*; (e) a leech (Hirudinea); (f) TS leech.

Hirudinea (leeches). Leeches have a clitellum, anterior and posterior suckers and no chaetae (Figure 9.2e,f). They are specialised predators, often ectoparasites, in fresh water, on land or sometimes in the sea. Oligochaeta and Hirudinea are now usually united as **Clitellata** (see 9.6, below).

9.3 What are the advantages of the coelom and of metamerism?

9.3.1 The coelom

A coelom is a body cavity surrounded by mesoderm and lined by a mesodermal epithelium, as was explained in Chapter 5. Such a cavity is useful in many ways. Separation of the gut and body wall enables each to move independently: the worm can bend without pushing food along the gut. A fluid-filled body cavity makes transport more effective and excretory organs, gonads and their products can be contained in the coelom. But probably the most important function of the coelom is to serve as a hydrostatic skeleton against which muscles can contract (see Chapter 5).

9.3.2 Metameric segmentation

Like the coelom, metameric segmentation is above all concerned with movement. The primary metameric structures are the body wall muscles, innervated in annelids by segmental nerves arising from a chain of segmental nerve ganglia. Segmentation allows control of particular portions of muscle, enabling each to contract independently of its neighbour. When each muscle contracts in succession, a wave of contraction passes along the worm and can set up lateral undulations or peristaltic waves that can be used for burrowing. First a 'penetration anchor' makes and enters the burrow, and then a 'terminal anchor' takes hold and pulls in the rest of the worm (Figure 9.3a). Such a burrowing sequence is common in unsegmented animals also, but is very much faster and more efficient where fluid localised by septa in a front coelomic compartment can build up considerable hydrostatic pressure, as in earthworms. Burrowers in soft sand, such as lugworms, by contrast lose some septa and move coelomic fluid forward in bulk.

9.3.3 Methods of locomotion

The slow walking of a typical polychaete such as *Nereis* is achieved by stepping with parapodia, their chaetae increasing traction with the substratum. Fast walking however uses the parapodia much less. The body is thrown into horizontal waves by the successive contraction and relaxation of the main body wall muscles alternately on the two sides; the lateral components of the force generated will cancel out and the worm moves forward by pressing backward on its environment (Figure 5.1a,b). When the polychaete changes from fast walking to swimming, the very rapid wave illustrates the activity possible for a segmented coelomate worm.

Polychaete lateral undulation differs from wave propulsion in a nematode, eel or snake in that there are projecting parapodia. They do not contribute much directly, but their presence, spread out on the outside and clustered on the inside of each bend (Figure 5.1b) means that the propulsive force is exerted on the outside convex surface rather than on the inside concave one. The wave of contraction has then to progress from the back to the front of the worm, unlike the front to back wave of a nematode, eel or snake, and the polychaete brain has to initiate locomotion at the far end of the body. A polychaete with all parapodia removed swims backwards.

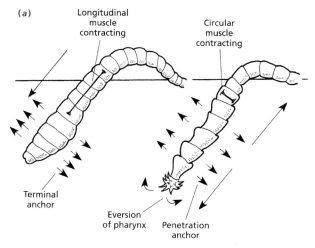

(a) Longitudinal muscle contracting / Circular muscle contracting / Terminal anchor / Eversion of pharynx / Penetration anchor

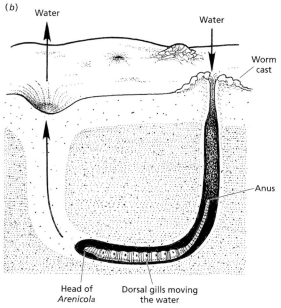

(b) Water / Water / Worm cast / Anus / Head of Arenicola / Dorsal gills moving the water

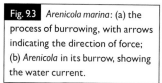

Fig. 9.3 *Arenicola marina*: (a) the process of burrowing, with arrows indicating the direction of force; (b) *Arenicola* in its burrow, showing the water current.

Oligochaetes use the alternating contraction of circular and longitudinal muscles not to undulate but for peristalsis (Figure 5.1a). Circular muscle contraction enables an earthworm to extend a longer thinner anterior portion while chaetae grip the ground in the portion just behind, made shorter and fatter by contraction of longitudinal muscles. The anterior circular muscles then relax and the wave of circular contraction passes backward along the worm. Leeches may 'loop' by attaching and detaching their suckers, or may swim by up-and-down waves propagated from the anterior end. To do this they need to flatten and stiffen their bodies by contraction of dorsoventral muscles, so that they can flex their muscles without the body shortening. The control of locomotion is more advanced than in other annelids, with a more condensed nervous system and more centralised functions.

9.4 How does a coelom introduce complexity?

The coelom both allows increase in size and activity and introduces a barrier between the outer body wall and the gut. A transport system therefore becomes essential, to supply the inner tissues with oxygen and the outer tissues with food.

9.4.1 The blood system

A small part of the primary body cavity or 'blastocoel' (see Box 5.2) of an annelid persists during development, becomes filled with blood and forms a closed system of vessels. General contractility of the dorsal longitudinal vessel pushes the blood forward, and then through a varying number of connecting vessels to flow posteriorly in the ventral vessel beneath the gut. While there is no true heart, there may be a number of blind ending contractile vessels that assist blood flow. There may be further vessels, for example to the parapodia in polychaetes. Such transverse vessels require the support of segmental septa (it is uncertain which of these structures came first and necessitated the other).

Leeches have very different circulatory systems. The coelom becomes packed with connective 'botryoidal' tissue, leaving coelomic channels in which fluid circulates. The blood system then becomes much reduced or is absent.

Box 9.2 | Respiration

Respiration is the oxidation of organic food with the release of energy. It is most commonly aerobic (using atmospheric oxygen) and has three stages:

1. **Gaseous exchange.** Oxygen is taken in and carbon dioxide is given off at the respiratory surface, often assisted by ventilation movements.
2. **Transport of oxygen** from the respiratory organ to the cells, usually in the blood, and of carbon dioxide in the reverse direction.
3. **Within each cell** the biochemical processes are remarkably similar in all animals. Sugars are assembled and broken down by glycolysis to pyruvate, releasing energy that is incorporated into ATP or lost as heat. This is the **anaerobic** stage of respiration: in the absence of oxygen the pyruvate formed is converted to lactic acid. This incomplete breakdown may be the sole stage of cellular respiration in some animals or some circumstances, e.g. when muscles build up 'oxygen debt'. Most animals, however, oxidise pyruvate further (in the 'Krebs' or 'tricarboxylic acid' cycle), using molecular oxygen as the final hydrogen acceptor. This **aerobic respiration** makes much more energy available.

The overall equation is then

$$C_6H_{12}O_6 + 6O_2 = 6CO_2 + 6H_2O$$

Aerial and aquatic respiration

Oxygen may be obtained from air or from solution in water.

Advantages of respiring in air

1. The oxygen content of air is about 30 times that of water: air is 21% oxygen, while a litre of fresh water at 15 °C will not contain more than 7.0 ml oxygen.
2. Oxygen diffuses 10 000 times faster in air than in water.
3. Air, being less dense and viscous, moves faster than water over the respiratory surface.

Advantages of respiring in water

1. The respiratory surface is supported by the water and will not dry out.
2. Carbon dioxide removal is very much easier: it is readily soluble, and fresh water can hold about 3 volumes % CO_2 while air contains only 0.03 volumes %.

Oxygen tension

The availability of oxygen in solution is measured in pressure rather than volume. The 'partial pressure' of oxygen (that part of the total which is due to oxygen) is called the oxygen tension. Atmospheric pressure at sea level is measured in a barometer as 760 millimetres of mercury (mm Hg: in SI units, 101.3 kilopascals, kPa). Air is about 21% oxygen, so the oxygen tension in well-aerated sea water is approximately $0.21 \times 760 = 160$ mm Hg (21.3 kPa).

Fresh water will in general be better oxygenated than the sea, because oxygen is less soluble in the presence of salts.

Clearly, on balance air is the easier medium for oxygen uptake, so that aquatic respiration will require a relatively large respiratory surface.

In the sea a large surface can readily be provided.

In fresh water, however, there is an osmotic gradient across the surface (see Box 7.1), and increased permeability to oxygen (and therefore also to water) will increase the energy requirement for osmoregulation.

Respiratory surfaces

The whole body surface

The surface of the body may be permeable to respiratory gases in marine animals, for example sponges, cnidarians, acoelomate worms, many small crustaceans and marine eggs and larvae.

Gills

The gills are thin-walled permeable extensions of the surface of an aquatic animal. External gills may be ventilated by water moving past the animal or by the animal moving through the water: ventilation of internal gills may require more energy. Gills and structures serving as gills include the parapodial lobes of free-swimming polychaetes, the tentacles (or separate gills) of tube-living polychaetes, the ctenidia of most molluscs and the mantle folds of limpets, all or parts of crustacean appendages and the tube feet of echinoderms.

Diagram to show the basic difference between (a) a gill and (b) a lung.

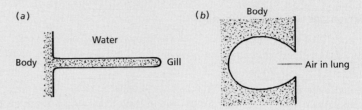

Lungs

The lungs are internal air-filled sacs with an opening to the outside world and blood closely applied to their outer walls. In invertebrate lungs gaseous exchange is by diffusion without ventilation movements, as in snails and slugs.

Tracheal systems

In insects and some other terrestrial arthropods, the tracheal systems consist of air-filled tubes branching all over the body and penetrating between the cells, providing extremely efficient aerial respiration for a small animal (see Chapter 15).

Respiration rate

The rate of respiration will vary according to the nature and circumstances of the invertebrate. It also will depend on body size, activity and ambient temperature. Respiration rate is usually equated with **metabolic rate**, the rate of the sum total of the activity of the animal. In aerobic animals metabolic rate can be measured by oxygen consumption per unit time; an alternative measure, applicable also to anaerobes, is the rate of heat production. Endothermic ('warm blooded') vertebrates maintain a constant 'basal metabolic rate' at rest, but ectotherms (including all invertebrates) do not: they have the advantage of being able to 'shut down', economising on water and energy, in adverse conditions. In those invertebrates where (for a given temperature) a resting maintenance metabolic rate can be measured, it is greater in larger animals but not scaled up 1:1 with body mass. As in vertebrates, resting metabolic rate tends to be proportional to body mass to the power 0.75.

9.4.2   Respiration

Permeability of the general outer surface is insufficient for most annelids to absorb enough oxygen, whether they are actively moving or living in a burrow or tube. Gills are required, such as parapodial lobes with a large surface area and internal blood supply. Tube-living polychaetes may need separate gills if the tentacles collect food from sand.

Respiratory pigments are now necessary (see Box 9.3), either in the blood to increase its oxygen carrying capacity or in tissues to facilitate diffusion. Haemoglobin is the commonest respiratory pigment, occurring in a very wide range of animals. It occurs in many annelids, but so also do two other rare iron-based pigments, haemerythrin and chlorocruorin, and the copper-based haemocyanin: no other phylum has so great a range.

Box 9.3 | Transport systems

The need for a transport system

Diffusion

Diffusion, the movement of a substance from a higher to a lower concentration of that substance, is not a sufficient means of communication between the parts of any but the smallest animals. Planarians, for example, rely on diffusion for their oxygen supply and typically consume about 0.1 to 0.2 ml oxygen per gram per hour at 15 °C. Calculations based on the rate of diffusion suggests that planarians should be no more than 0.5 mm thick. [They do not all obey. At this point biology takes over from physics: most planarians respire anaerobically part of the time, and much of the centre of the worm is not tissue but food in the gut. The general point, however, is valid.]

The transport or 'vascular' system

The commonest transport system is blood, a fluid tissue (i.e. it contains cells) either in closed vessels or in a haemocoelic cavity. Its function is to transport respiratory gases, food, excretory matter, hormones and many other substances. The assumption that the earliest blood systems evolved primarily for oxygen transport may be mistaken: nemertines, for example (see Chapter 7), have a blood system lying deep in the body and not mainly concerned with oxygen transport. Nemertines, unlike planarians, have a gut opening at both ends with one-way movement of food; this at once demands a transport system if food is to reach all parts of the body.

Even in planarians, where the mouth is the only opening, food would not reach all tissues but for the considerable branching of the gut.

Transport of oxygen and carbon dioxide

Respiratory pigments

Pigments that combine reversibly with oxygen are in most animals a necessary aid to respiration. By far the commonest is haemoglobin (Hb), found at least in some members of almost every phylum. Structurally a protein containing iron and haem, it is related to the cytochrome respiratory enzymes used in cellular respiration in all animals. Haemoglobin may occur in solution or in corpuscles, where it can become very much more concentrated without raising the osmotic pressure of the blood. Primarily it transports oxygen, but it may be an oxygen store or, often as a simpler form of the molecule called 'myoglobin', it may facilitate oxygen diffusion. Haemocyanin is a copper-based pigment occurring in many arthropods and molluscs. It has a lower oxygen-carrying capacity than haemoglobin and cannot be contained in corpuscles. Other respiratory pigments such as haemerythrin and chlorocruorin occur, especially in annelids, which have an especially wide range of respiratory pigments.

(a) Oxygen dissociation curves at the respiratory surface (A) and the internal tissues (B). (b) Oxygen dissociation curves for *Arenicola* within its burrow (C), *Lumbricus* at 7 °C (D), *Lumbricus* at 20 °C (E).

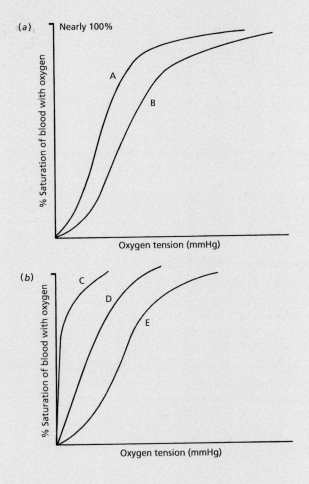

Carbon dioxide

Carbon dioxide is also transported in reversible combination with haemoglobin, as bicarbonate (HCO_3)$^-$ ions or in physical solution.

Oxygen dissociation curves

Graphs showing oxygen dissociation (see diagram) express the relationship between the amount of oxygen available and the amount taken up by the blood. They are informative about the haemoglobins in different animals or in different circumstances. The characteristic sigmoid curve may be shifted either to the left, when oxygen can more easily be taken up, or to the right, when oxygen is more easily given off in the tissues. The 'Bohr shift' exemplifies these two positions of the curve within most animals: in diagram (a) curve A represents conditions at the respiratory surface and the shift to the right (to curve B) is caused by the higher CO_2 concentration (causing greater acidity) at the tissues, enabling more oxygen to be given off from the haemoglobin molecule. Animals living in places with low oxygen tensions, where ease of oxygen uptake is paramount, have curves well to the left – see diagram (b), curve C, for the polychaete *Arenicola* in its burrow. Other factors affecting the balance of advantage between ease of uptake and ease of release of oxygen are temperature (see curves D and E for the earthworm *Lumbricus*) and high altitude, where there is less total oxygen due to low atmospheric pressure and the curve moves to the left.

Countercurrent systems

Where the direction of flow is opposite in closely apposed channels, countercurrent systems increase the uptake of oxygen at respiratory surfaces. With countercurrent flow there is a concentration gradient of oxygen (or any other substance, or heat) along the whole surface in contact, allowing very much more diffusion than can occur in parallel ('co-current') flow, where there is a high initial gradient progressively decreasing – see diagrams (c) and (d).

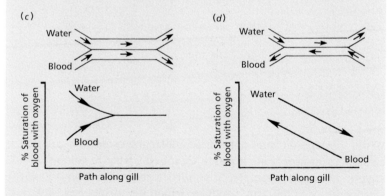

(c) Co-current and
(d) countercurrent flow of water and blood at a gill, showing gradients of oxygen content as the water passes along the gill.

The transport systems of different invertebrates

No transport system

No transport system occurs in sponges or in most cnidarians, where every cell is close to the ambient medium. None is present in platyhelminths, acoelomorphs, gnathostomula, gastrotrichs, nematodes (where fluid under high pressure is moved in the body cavity), nematomorphs, rotifers, acanthocephala, kinorhynchs, priapulids (though there are cells containing haemerythrin in the body cavity), entoprocts, sipuncula, bryozoa or chaetognaths.

Seawater transport systems

Seawater channels occur in large jellyfish (Scyphozoa). Although both cell layers are in contact with the sea for respiratory exchange, very thick mesoglea may separate the mouth from the ectodermal structures. Food in sea water circulates in canals through the jelly: straight radial ones from mouth to margin and much-branching canals where water returns inward, as can often be seen through the translucent jelly (see Figure 4.5). Echinoderms have no blood system but a very well developed coelomic water vascular system, where sea water is drawn in, circulates and is let out.

Closed blood systems

Systems where blood circulates (forward dorsally and backward ventrally, except in chordates, where it is the reverse) in narrow vessels, propelled by hearts or general contractility of the main vessel, occur in annelids (see Figure 9.2), phorona and chordates; nemertines have a differently derived closed system, and in the very active cephalopod molluscs the open haemocoel has become canalised into vessels, enabling faster circulation.

Circulation in channels of the coelom

Coelomic circulation occurs in leeches, where the coelom is invaded by connective tissue or 'botryoidal' tissue, leaving a system of sinuses and channels where fluid circulates, propelled by contractions of the main lateral vessels. The blood system is much reduced or absent.

Haemocoels

Haemocoels, open blood systems, occur in molluscs, brachiopods and arthropods (see Figure 12.5). Insects have a tracheal system that brings oxygen to every cell. The blood is not involved in respiration and (except in a few larvae) there are no respiratory pigments.

9.4.3 Excretion

While most polychaetes are osmoconformers, osmotic regulation is an important aspect of excretion in annelids able to colonise dilute environments. For example, *Nereis diversicolor* in estuaries and earthworms in wet soil can produce hypotonic urine (see Box 7.1). Nitrogenous waste may be excreted largely in the form of ammonia, or where water is less freely available specialised cells (in oligochaetes) or botryoidal cells (in leeches) may combine ammonia with carbon dioxide to make urea.

The excretory organs in larvae and some adult polychaetes (and possibly in ancestral annelids) are 'protonephridia', tubes with blind ends (usually flame cells, see Box 9.1) where cilia-mediated filtration occurs. Most polychaetes have 'metanephridia', ducts opening from the coelom by ciliated funnels. The tubes may lead separately to the outside but more often open into the mesodermal gonadal ducts occurring in most segments.

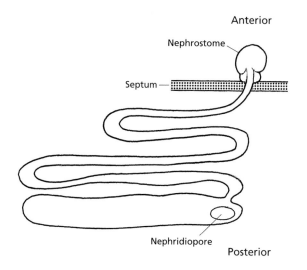

Anterior

Fig. 9.4 The metanephridium of an earthworm.

Oligochaetes and leeches have segmental metanephridia (Figure 9.4), with gonads and their ducts restricted to a few anterior segments.

9.5 How do annelids reproduce and feed?

Polychaetes are by far the largest class of annelids, showing remarkably great diversity, while the oligochaetes (and still more the leeches) are more uniform. The division between polychaetes and oligochaetes is at first sight extremely tidy, with a habitat difference typically related to reproduction, as follows.

9.5.1 Reproduction

Polychaetes are marine and have separate sexes, external sperm transfer and external fertilisation. The eggs are shed through the segmental organs into the sea, where they develop into planktonic 'trochophore' larvae (Figure 9.5). Oligochaetes, in fresh water or in soil, are cross-fertilised hermaphrodites with no external sperm transfer: the yolky eggs are fertilised within the worm or in the cocoon. The clitellum, situated near the female pores, secretes the cocoon into which (before or after fertilisation) the eggs are shed, and where they develop directly with no larval stage. Leeches in any habitat reproduce like oligochaetes, with direct development in a cocoon.

Generalisation provides a useful framework but (as so often in biology) breaks down on account of the versatility of natural selection. Polychaetes are not all marine but have a few freshwater species, and rather more living under damp rotten logs on land; oligochaetes have a number of marine species. While the

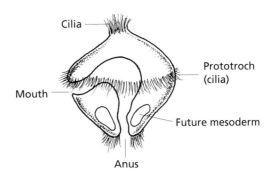

Cilia

Prototroch (cilia)

Mouth

Future mesoderm

Anus

generalisations about reproduction and development in oligochaetes may hold, polychaetes vary widely. The female may collect and store sperm, for example, or may be viviparous; the trochophore larva may be modified or suppressed where development proceeds in a yolky egg. Asexual reproduction (with corresponding regenerative powers) may occur in polychaetes by budding, or may be combined with the sexual process as when the back end of a *Nereis* breaks off and becomes transformed into a pelagic 'heteronereis': like other pelagic polychaetes, it then floats due to its extended surface and has very large eyes. Gamete release may be synchronised in a swarm of sexually mature polychaetes.

9.5.2 Feeding

Polychaetes freely crawling or swimming in the sea are usually brightly coloured active predators, with biting jaws borne on an eversible pharynx. Most are carnivores, but herbivores, detritivores and omnivores are known. There are very many genera, mostly with names inspired by Greek nymphs and goddesses. *Nereis* (Figure 9.2a) is a familiar 'ragworm', as are *Nephtys* and *Phylloduce*; the Syllidae are slender with feathery parapodia, the Glyceridae have balloon-like pharynges, the Eunicidae have especially strong jaws and the scale worms such as *Aphrodite* have gill covers all down their backs. There are some 25 families, formerly grouped as 'Errantia', but they are not all closely related. The arrival of arthropod predators was no doubt a stimulus for rapid evolution among relatively defenceless worms; many times separately they took to living in burrows as sedimentary or detritus feeders. External structures became reduced but muscles remained well developed and gills were needed. Examples include *Arenicola* the lugworm (Figure 9.3b), responsible for worm casts on sandy and muddy beaches, and *Chaetopterus*, with remarkable regional differentiation of modified appendages which move water through its U-tube. Sessile tubicolous polychaetes, secreting the tube in which they live, are highly modified, with U-shaped guts. Well developed nerves and muscles are important not for locomotion but for very fast retraction into the tube. The head becomes a crown of tentacles that may stretch over the sand and pick up particles, as in Terebellidae, or be held up, enabling ciliary

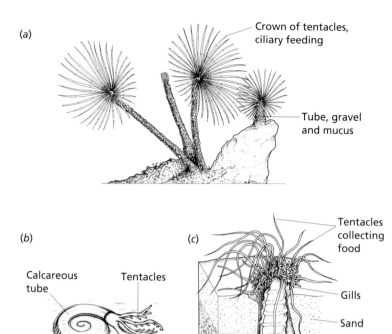

(a)

Crown of tentacles,
ciliary feeding

Tube, gravel
and mucus

Fig. 9.6 Drawings of tubicolous
polychaetes: (a) *Sabella*;
(b) *Spirorbis*; (c) *Amphitrite*.

(b)

Calcareous
tube

Tentacles

Tube
stopper

(c)

Tentacles
collecting
food

Gills

Sand

feeding as in the fan worm *Sabella* (Figure 9.6a) and in the Serpulidae, makers of the calcareous tubes commonly found on rocks or seaweed (Figure 9.6b).

Diversity in feeding is slight among oligochaetes, which have a non-eversible pharynx and eat plant detritus, and even slighter among the leeches. Such is their morphological uniformity that Hirudinea all have 33 segments and their embryology is almost as rigidly determined by cell lineage as in nematodes.

9.6 How are annelids related to each other?

The relationship between the three main classes of annelid is clear only in that the Clitellata are distinct, whatever their origin. Within this group, there can no longer be a clear demarcation between classes: the highly specialised Hirudinea (leeches) arose, probably more than once, from among the Oligochaetes. It seems that the Clitellata were originally terrestrial worms, many of which secondarily came to live in water.

The origin and composition of Polychaeta is very far from clear. Many 'worms' once thought to belong to separate phyla are now recognised as modified polychaetes. First the 'Archiannelida', a collection of small and therefore simple annelids once thought to be ancestral, were recognised as being neither united nor primitive but aberrant polychaetes from various families. Then the Echiura and possibly also Sipuncula (see Chapter 5) were seen to constitute a group of polychaetes which have as adults lost their segmentation and the power of locomotion. Now the Pogonophora are believed, with support from molecular evidence, to have arisen within the polychaetes: they are now called 'Siboglinidae' and defined as the first group of polychaetes to have the gut occluded by expanded endoderm filled with chemosynthetic bacteria. They may have arisen close to *Sabella*.

Relationships among the many orders of polychaetes are very hard to elucidate: clearly the protection of burrows and tubes was achieved many times separately by different ancestors.

There has long been controversy about the size and nature of the annelid ancestor; now the question is, did the polychaetes (and therefore the annelids as a whole) have a common ancestor at all? To apply the language of cladistics, polychaetes may not be 'paraphyletic' (an assemblage including some but not all members of a monophyletic group) but 'polyphyletic' (a group of animals with more than one ancestor). Mercifully we have insufficient evidence to pursue these findings to a conclusion, and we continue at present to talk about 'Polychaeta' and 'Annelida' as we always have done.

9.7 How are annelids related to other phyla?

The first step towards an answer is clear: annelid early development is unambiguously protostome (see Box 5.2 and Chapter 19). Cleavage is spiral, the blastopore becomes the mouth and the mesoderm is early set aside, later becoming blocks that form segments and split to make the coelom. Annelids are traditionally placed near the base of the protostome radiation and close to both the molluscs and the arthropods; more recent morphological assessment has maintained the link with molluscs but has denied that the common possession of metameric segmentation united annelids and arthropods. Molecular evidence supports the complete separation of annelids and arthropods and the placing of annelids in a group with other soft-bodied worms and with molluscs, but the relationships within this group remain uncertain.

Molecular evidence is an exciting new guide to relationships, but it cannot in isolation tell us anything about the direction of evolution, nor can it replace the need to study the animals themselves. Observing the rapid and well-controlled locomotion of many annelids makes it hard to believe that secondary loss of a coelom could have been advantageous for an active predator. Similarly, it seems most improbable that nemertines evolved from animals with coelomic body cavities: it is far more likely that localised structures such as the rhynchocoel evolved convergently. Annelids show us that metameric segmentation, like the coelom, can be a great functional asset, and we should not be surprised that it almost certainly evolved several times separately.

Phylogenies uniting animals because they share such clearly advantageous characters may be very misleading: the likelihood of convergence should not be underestimated.

Chapter 10

Mollusca: general and Gastropoda

'Life in a shell' is the molluscan theme; it sounds very restricted. Yet Mollusca constitute a remarkably large and wide-ranging phylum of at least 100 000 species, mostly marine but including many freshwater species and about 10 000 land snails. The shell is used in a variety of ways, or discarded altogether. The molluscan body plan is unique, and can be traced in animals as different as snails, scallops and squids, functioning over an unusually large size range (from a few millimetres to 15 metres). This account first introduces the distinctive body plan and explains how the shell may be used and then, for each of the main groups, indicates how the basic design has become modified and how so many different molluscs manage to make a living.

10.1   What is the basic molluscan body plan?

Molluscan groups which (at least primitively) have shells share a unique body plan (Figure 10.1a) The internal organs ('viscera') are contained between a ventral muscular creeping foot and a dorsal calcareous shell secreted by an underlying epidermal covering, the mantle. The mouth opens anteriorly into a buccal cavity with a small sac containing the radula, a rasping tongue covered in teeth made of chitin, a distinctive feature unique to molluscs. The mantle tends to overhang the body as a pair of double-walled folds, leaving posteriorly a space filled with ambient water. This mantle cavity contains the 'ctenidium' (morphological term) or 'gill' (a name indicating its respiratory function). The gut bears a digestive gland, often the site of both enzyme secretion and food absorption: the anus opens into the mantle cavity. The body is unsegmented and the main body cavity is a haemocoel. There is also a very much restricted coelom, consisting of a space round the heart and the cavities of the excretory and reproductive organs. The brain primitively is a simple ring round the oesophagus, with two longitudinal nerve cords. Sensory tentacles are usually present and there may be eyes. Development is by spiral cleavage and coelom formation is schizocoelic, i.e. molluscs are typical protostomes (see Box 5.2). There may be a trochophore-type larva (as in polychaetes) followed by a 'veliger' with a shell rudiment.

10.2   How can such an animal function?

10.2.1   Feeding: the radula

The radula is unique to molluscs, and occurs in all groups except bivalves, which have abandoned feeding with a rasping tongue. The radula consists of chitinous teeth stretched over a supporting belt,

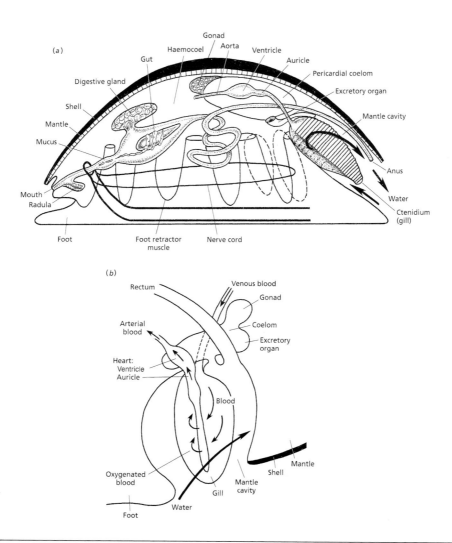

Fig. 10.1 Diagrams indicating basic molluscan structure and function: (a) diagram of basic body plan; (b) diagram of a primitive gastropod to show the water flow over the gill, the blood circulation and the relationship of the heart and excretory organs.

pointing backwards. The teeth are in horizontal rows, shapes varying according to their position, and are continuously worn away and replaced as the supporting belt moves forward. The radula is primitively a scraping organ used when collecting algae and other small food organisms from hard surfaces, but has become adapted for a wide range of diets (see below).

10.2.2 Respiration: the ctenidium (gill)

Typically a pair of ctenidia hang in the ciliated mantle cavity. Each ctenidium has a central axis with a lamella on each side, and is covered in cilia. Water carrying oxygen is driven by ciliary action to enter the mantle cavity ventrally, make a U-turn over the gill and exit dorsoposteriorly: it therefore flows over the gill from front to back. The blood inside the gill flows across in the opposite direction, as shown in Figure 10.1b. The blood, with its affinity for oxygen increased by haemocyanin in solution, receives oxygen by diffusion from the water stream and is collected into a central vessel. Sensory organs beside the gill monitor the quality of the incoming water; the exhalant water stream removes waste and any genital products from the mantle cavity.

10.2.3 Blood circulation: the haemocoel

Oxygenated (arterial) blood from the ctenidium is pulled into the contracting heart (Figure 10.1b), and from the ventricle is discharged into the open space of the haemocoel, where it bathes the tissues. Circulation is slow. Deoxygenated (venous) blood enters the base of the ctenidium, becomes oxygenated as it passes through the small vessels in the lamellae, then continues by way of the larger central vessel into the ventricle.

10.2.4 Excretion: the coelom

The excretory organs of molluscs are metanephridia (see Box 9.1). Waste collected from the haemocoel filters into the pericardial coelom: this is the largest part of the molluscan coelom (Figure 10.1b), between the heart and its enclosing organ, the pericardium. Waste is then discharged into the mantle cavity through a duct which may also carry the gametes.

10.2.5 Locomotion: the foot

The foot is strongly muscular. The primitive ventral creeping foot moves the animal forward by waves of muscle contraction working against fluid pressure from neighbouring cells and from the haemocoelic spaces within the foot. Production of mucus is a necessary aid to locomotion. The muscle waves may either be 'direct' (in the direction of locomotion, as in polychaetes) or 'retrograde' (in the opposite direction, as in nematodes) according to the

(i)

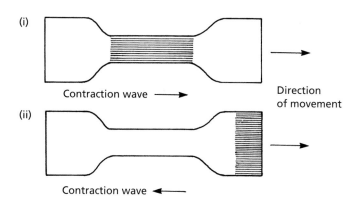

Contraction wave ———▶

Direction
of movement

(ii)

Contraction wave ◀———

Fig. 10.2 The locomotor muscular waves over the molluscan foot (viewed from above): (i) direct waves, in the direction of motion; (ii) retrograde waves, in the opposite direction to the motion. Shaded areas show points of attachment to the substratum.

stage in the contraction cycle at which the foot is stationary on the ground (Figure 10.2).

Some very small molluscs are instead propelled by cilia on the foot; larvae (Figure 10.3a) also use ciliary locomotion.

10.3   What is the shell and how may it be used?

The shell consists of two or more layers of calcium carbonate deposited within a framework of a protein called conchiolin, and is covered externally by a different protein layer called the periostracum. The calcium carbonate is derived from calcium bicarbonate in solution and laid down as crystals of calcite or aragonite. The shell is secreted by the mantle, mainly at the periphery where the outermost layers are deposited, but there may be some thickening over the general mantle surface. Shells are stiffer and weaker than our bones: stiffer because of the high proportion of inorganic crystals and weaker because there is less of any strengthening protein. The innermost 'nacreous' (mother o' pearl) layer has sheets of aragonite all orientated parallel to the surface, and is stronger than the outer 'prismatic' layer made of crystals. The horny outer periostracum protects the calcium carbonate from carbonic acid and other agents of damage.

Unlike the arthropod exoskeleton, the molluscan shell does not fit the body very closely and is not moulted as the animal grows. It may support the body but does not provide its framework. It is not jointed and does not primarily serve to attach muscles. Primitively the shell was probably a shield protecting the soft-bodied mollusc against predators, mechanical damage and (out of the sea) desiccation. It may be modified into protective housing that can completely enclose or be pulled down over the soft body, or the body can be

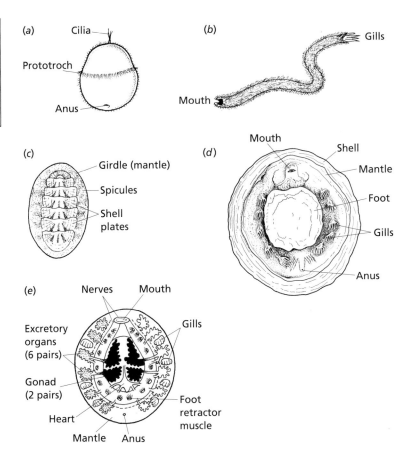

Fig. 10.3 (a) A gastropod trochophore; (b) an aplacophoran; (c) a chiton (polyplacophoran); (d) *Neopilina*, a monoplacophoran, in ventral view; (e) a diagram of *Neopilina* showing the multiplication of organs.

pulled into the shell, by contraction of pedal retractor muscles originating on the inner shell surface and attached to the foot. The prevalence of molluscs in the intertidal region and of snails on land shows that the protection afforded by the shell has been very important.

The shell has many other functions in different molluscs. Muscles may be attached to it, for example those effecting torsion in gastropods (see below) and the adductors holding together the two halves of bivalves. It can be used for burrowing or boring. It may form a keel in pelagic molluscs or it may contain gas and assist buoyancy. Yet all groups except bivalves have a tendency for evolutionary loss of the shell. Loss of protection is balanced by the gain in mobility and lightness, and in different circumstances either advantage may be paramount. For example, snails are well protected above ground. Slugs can obtain comparable protection from desiccation and predators only by burrowing beneath the surface, as their shape and lack of shell allows. Snails are strikingly plentiful and diverse on calcareous soils; where there is less calcium available, slugs may be commoner.

10.4   What are the main groups of molluscs?

Figures 10.3 to 10.6 show examples of those discussed in this chapter:

Aculifera, with spicules. Divided into two classes:
   Aplacophora: worm-like molluscs, without shells.
   Chaetodermomorpha, with no foot; burrowing.
   Neomeniomorpha, with ridge representing foot; creeping.
   Polyplacophora: chitons, with eight dorsal shell plates and spicules in the surrounding mantle. Ventral creeping foot.
Conchifera, with shells, not spicules; show typical molluscan body plan. Divided into five classes:
   Monoplacophora: shell a single cap or cone.
   Gastropoda: shell single, often coiled. Torsion during development brings mantle cavity to the front.
      Prosobranchiata: gill in front. A very large diverse group including limpets, winkles, whelks. Mostly marine.
      Opisthobranchiata: gill behind, due to detorsion associated with loss of shell. Includes sea slugs.
      Pulmonata: mostly terrestrial. The gill is lost and the mantle cavity becomes an air-filled 'lung'. Land snails and slugs.
   Bivalvia: shell as two lateral valves, hinged at the top, enclosing the body. Typically sedentary. Clams, mussels, oysters, scallops.
   Scaphopoda: shell tubular, opening at both ends: a small and uniform marine class.
   Cephalopoda: actively carnivorous marine molluscs.
      Ammonoidea, with coiled external shell. Extinct ammonites, common as fossils.
      Nautiloidea, with coiled external shell, including present-day forms.
      Coleoidea: much modified molluscs with highly developed brains, sense organs and nervous systems, swimming fast by jet propulsion. The shell is either internal and reduced or lost entirely. Cuttlefish, squids, octopus.

10.5   What are the Aculifera?

The two classes of molluscs which have spicules are the Aplacophora and the Polyplacophora.

Aplacophora are worm-like forms with spicules but no shell (Figure 10.3b). They possess a radula, which diagnoses them as molluscs. They are small, mostly 1–5 mm in length. Some 288 species are known, mostly living in the deep sea, where there are probably many more species undescribed. Several features of these animals may be primitive for molluscs – the phylogeny is controversial – but they cannot be assumed to represent ancestral forms; nor can the two groups, chaetodermomorphs and neomeniomorphs, be assumed to have common ancestry.

Polyplacophora are quite unlike other molluscs in having eight overlapping dorsal shell plates, surrounded by a spicule-bearing 'girdle' of the mantle (Figure 10.3c). Each plate is penetrated by living tissue with sensory structures and nerves, and has a separate pedal retractor muscle. Their embryological formation also is different: the plates are laid down as rods in dorsal horizontal grooves.

Polyplacophora are known from the early Cambrian (in Australia) and there are about 800 species of living chitons, over half living in shallow waters. They are common in the intertidal region (in the southern hemisphere the shores are strikingly rich in large chitons). They creep over rocks, grazing mainly on algae. Small gills are multiplied in rows within the mantle cavity, which extends on either side of the powerful foot, which by contraction of the pedal retractors can clamp the animal down on to the rock.

Polyplacophora may retain some primitive molluscan features, but they are not seen as ancestral to the phylum. They are widespread and numerous, a separately successful class of molluscs.

10.6 What is unusual about the Monoplacophora?

Monoplacophora, molluscs with the shell as a single cap or cone, are characterised by multiplication of their organs. They were known only as fossils until a deep-sea dredge in 1954 brought up *Neopilina* (Figure 10.3d,e). Although simple in structure it appeared to have two extensive dorsal coeloms and some structures were multiplied (five pairs of gills, six pairs of excretory organs, eight pairs of pedal retractors and two pairs of gonads). *Neopilina* was hailed as close to the ancestral mollusc, with a well-developed coelom and, although the numbers did not coincide, vestigial segmentation. Since then ten more species have been found, none longer than 30 mm, living as specialist detritus feeders in the ooze at the bottom of very deep seas. Further study revealed that the 'coelomic' cavities were linked to the pharynx, and the (varying) multiplication of some organs is now seen as a specialised feature.

Monoplacophora may be descendants of the stem group of shell-bearing molluscs, but their unusual features do not illuminate the evolutionary history of the phylum.

GASTROPODA

Gastropods are the largest and most diverse group of molluscs (some 50 000 species have been described). They occur at all levels of the sea, in fresh water and on land, with a remarkable range of feeding methods, even including internal parasitism. The impression is, 'You name it, some gastropod does it.' This account cannot do justice to such diversity; it can only indicate its basis.

10.7   How is the molluscan body plan modified in gastropods?

The head is usually well developed with eyes and sensory tentacles. The primitive ventral creeping foot is usually retained, and the 'visceral hump' includes the digestive gland and part of the gut. The shell is in one piece, enclosing the visceral hump and the mantle cavity, all as represented in Figures 10.1a and 10.4.

10.7.1   Shell coiling

The shell is often coiled. Development as an expanding cone would make the visceral hump large and unwieldy unless it were coiled, and the shell becomes coiled also.

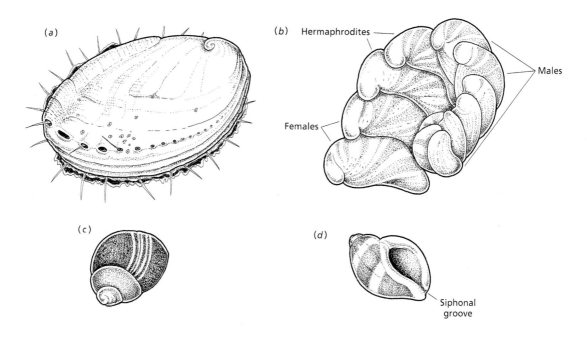

Fig. 10.4   Drawings of some prosobranch gastropods: (a) *Haliotis*, an ormer; (b) *Crepidula*, the slipper limpet, showing the gradation of the sexes; (c) *Littorina*, a periwinkle; (d) *Nucella*, the dog whelk.

10.7.2 Torsion: prosobranchs

The outstanding characteristic of gastropods, however, is torsion, a phenomenon entirely separate from coiling of the visceral hump and of the shell. At the veliger larval stage the visceral hump is twisted through 180 degrees, bringing the mantle cavity to the front (hence the name given to the basal group of gastropods, 'prosobranchs', Figure 10.5). The foot (except in pulmonates) carries an operculum which can close the opening when the body is withdrawn into the shell.

Torsion through 90 degrees is achieved in a few hours, by contraction of an asymmetrical muscle attached to the right side of the shell and the left side of the head and foot (Figure 10.5a,b). At the point of twisting the two main nerve cords are crossed, making a figure of 8. Torsion is completed more slowly, by asymmetrical growth of the larva.

What is the adaptive significance of this curious phenomenon? Possibly it helps the mollusc when attacked to draw the head into the shell before the back end of the foot, but that is not very convincing: a small larva is likely to be swallowed in one gulp. A more satisfactory theory concerns the water currents in the mantle cavity. Where the mantle cavity is posterior and overhung by the shell, it may be difficult to achieve a respiratory current flowing in the opposite direction to the current flowing past a forward-moving animal

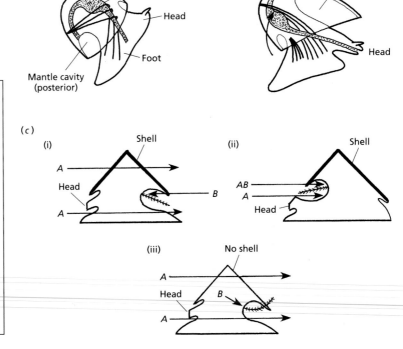

Fig. 10.5 Torsion in gastropods. Limpet veliger (advanced larva): (a) before torsion; (b) after torsion. Arrows show direction of twisting. (c) Diagrams of water currents (A, as the animal moves forward, and B, respiratory current) in (i) prosobranch without torsion, (ii) prosobranch with torsion (making the mantle cavity anterior), (iii) opisthobranch with no shell and no torsion. (d) Water currents in the mantle cavity of *Haliotis*. (e) Dorsal view of water currents in the mantle cavity of (i) a prosobranch with two gills and (ii) a prosobranch with one gill.

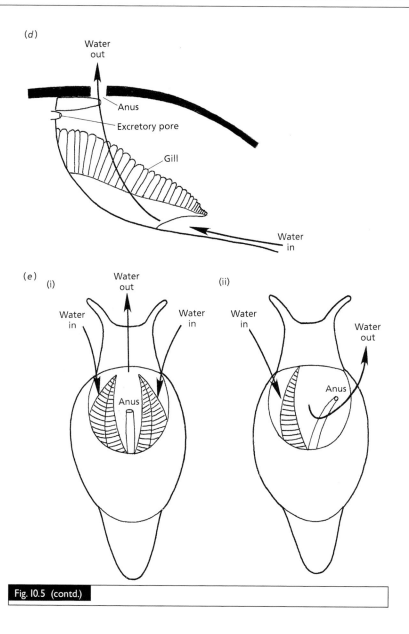

Fig. 10.5 (contd.)

(Figure 10.5c). After torsion this difficulty disappears, as fresh clean water flows readily into an anterior mantle cavity. A new problem arises, however: with the anus now anterior, waste will be deposited on the gill. This problem is reduced if the shell develops a hole or a series of holes: the anus can be retracted and the outflowing respiratory current will carry the faeces (Figure 10.5d). This condition obtains in the more primitive prosobranchs, e.g. the abalone (ormer) *Haliotis* (Figure 10.4a), the slit shell *Emarginula* and the keyhole limpet *Diodora*. True limpets that clamp tightly down on the rocks, for example *Patella*, do not have this problem, since the mantle cavity is greatly reduced and the ctenidia are entirely lost: respiratory exchange occurs in folds of the mantle edge ('pallial gills'). The vast majority of prosobranchs (including periwinkles (Figure 10.4c), cowries and conch shells) have lost one ctenidium and draw in water from the side, first over the single gill then out over the anus (Figure 10.5e). The most advanced prosobranchs are the carnivorous whelks and dog whelks (Figure 10.4d), also with a single ctenidium. They are recognisable by a groove at the shell opening, where a siphon is projected as they hunt for prey.

10.7.3   Opisthobranchs

The members of this group have untwisted: the mantle cavity is again at the back. That this process is truly a secondary 'detorsion' is shown by the anatomy of the nerve cords, by remnants of torsion in some species and by the retention of a single ctenidium. Detorsion has occurred many times in parallel, in combination with the loss of the shell (without a shell, clean water readily reaches a posterior mantle cavity, Figure 10.5c). Opisthobranchs include the sea hare *Aplysia* (Figure 10.6a), much used in neurobiology on account of the simple accessible nervous system and the large nerve cells, the sea butterflies, with or without shells, which swim by means of a modified foot, and the 'nudibranchs', entirely without shells, for example the sea slugs *Archidoris* and *Aeolidia* (Figure 10.6b,c).

10.7.4   Pulmonates

In this group the mantle cavity has become an air-filled lung (Figure 10.6d), and they may also show anatomical evidence of detorsion. The many freshwater pulmonates either carry an air store, frequently replenished at the surface, or, as in the freshwater snail *Planorbis*, they have secondarily developed gills.

10.8   How may gastropods feed?

Primitively, small particles of food are scraped off hard surfaces by the radula, as occurs in browsers and grazers such as limpets, top shells and periwinkles and many other intertidal prosobranchs. Small food particles are also collected in many other ways, mainly in the sea: detritus feeders extract decomposing organic material from the sea bottom; suspension feeders include those trapping food in mucus, such as the pelagic sea butterflies and ciliary feeders such as *Crepidula*, which uses enlarged ctenidia with many ciliary tracts (much as in bivalves).

Many gastropods, however, use the radula to obtain larger pieces of food, as herbivores, scavengers or carnivores. The radula may cut, pierce, grasp or even masticate the food and there may in addition be jaws. Herbivores include most pulmonates, opisthobranchs such as *Aplysia* and many freshwater prosobranchs. Carnivores include the

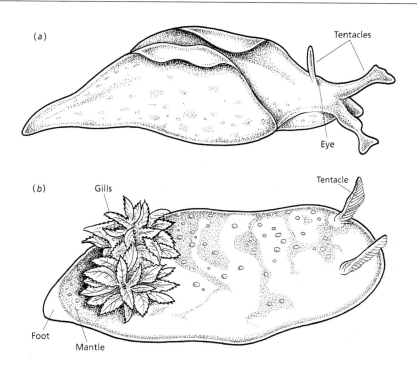

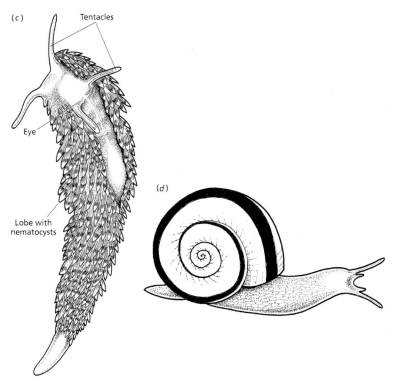

Fig. 10.6 Drawings of three opisthobranchs and one pulmonate: the opisthobranchs (a) *Aplysia*, the sea hare; (b) *Archidoris*; (c) *Aeolidia papillosa*; (d) *Helicella*, a pulmonate snail.

cowries browsing on ascidians, and neogastropods that use a proboscis to find and seize their prey. Of these, whelks use the foot to smother the victim or to wedge open the valves of a bivalve, *Urosalpinx* in oyster beds secretes acid which assists penetration of the oyster shell by a drill-like radula, cone shells have poisonous radula teeth and a bite may be fatal to humans. Pulmonates may eat earthworms or slugs: the slug *Testicella* seizes earthworms, grips them with the radula and swallows them whole.

The aeolid opisthobranchs (Figure 10.6c) have the remarkable ability to live and feed on hydroids, anemones or even corals; they may resemble them so closely that when feeding the slugs are very hard to see. They are not stung by the cnidarian nematocysts (see Chapter 4); they pass these stinging cells, undischarged, by way of the gut to their dorsal lobes (Figure 10.6c). Recent work shows that mucus from the aeolid inhibits nematocyst discharge. The mucus does not merely form a physical barrier: its composition varies according to the prey to inhibit the particular nematocysts of the species concerned. An aeolid can change its mucus within two weeks, on encountering a different species of anemone. In this way the mollusc has working stinging cells for its own protection, in place of a shell.

Ectoparasitic gastropods include *Thyca*, a prosobranch that has lost the radula, with the mouth modified into a sucker by which it attaches itself to a starfish and extracts soft tissue. Endoparasites occur in one prosobranch family, for example the worm-like *Enteroxenus*, without a shell, which lives in sea cucumbers and absorbs food all over its surface.

10.8.1   Digestion

Food is moved along the gut by the beating of cilia rather than by muscular contraction. Usually the stomach is the site of extracellular digestion, by enzymes from the salivary glands. The food is then sorted in the caecum: the fine particles pass to the main digestive gland, where there may be intracellular digestion as well as absorption. Heavy rejected particles pass to the intestine, being stored there when the animal is withdrawn into the shell, or voided at the anus.

10.9   Why are many gastropods hermaphrodites?

Adaptations for sexual reproduction in gastropods illustrate the balance of advantage between having separate sexes or having the organs of sexual reproduction in one body (i.e. being hermaphroditic). Animals with separate sexes have the economy of producing only one set of sexual organs and the opportunity for advantageous division of labour between a mobile male and a food-storing female. To have separate sexes is by far the commoner reproductive strategy, especially in the sea. Hermaphrodites may have the advantage, however, where the opportunities for animals to meet is reduced, since every meeting is a potential mating. Cross-fertilisation is advantageous in that it introduces genetic variation, but if no meetings occur, self-fertilisation may be possible for a hermaphrodite. Sessile animals are usually hermaphroditic. Colonisation of land or even fresh water, especially in isolated or newly colonised habitats, may separate individuals, in particular if they are slow-moving like molluscs. Further, sense organs evolved in the sea may not work so well for finding mates on land.

Most marine prosobranchs have separate sexes. Primitively the gonads have no separate ducts, but shed their products into the pericardial coelom or directly to one of the excretory ducts, which carry them to the mantle cavity. Eggs and sperm are washed out to sea and fertilisation is external. More advanced prosobranchs have gonadal ducts and a penis in the male, allowing internal fertilisation. Some species are parthenogenetic: they have no males and the eggs develop without being fertilised — for example the common small freshwater prosobranch *Potamopyrgus jenkinsi*, an immigrant from New Zealand. A few prosobranchs are hermaphroditic (including the sessile limpets) and most hermaphrodites are protandrous (i.e. the male organs develop first). A striking example is the slipper limpet *Crepidula fornicata* (Figure 10.4b), where individuals live stacked up in a pile ('fornicata' means 'arch-forming'). All the young are at first solitary and male: they become hermaphroditic and then female. Other young males arrive and make a pile. They begin to feminise, but the first female produces a hormone (called a 'pheromone' since it acts outside her body) that maintains the masculinity of any male in her vicinity. In a typical pile the bottom (oldest) animals will be female, the next one or two hermaphrodite and the youngest male. All the females can then be cross-fertilised.

Opisthobranchs and pulmonates are all hermaphrodites. Opisthobranchs are slow-moving animals with specialised feeding habitats, giving them a patchy distribution. The land-living pulmonates are cross-fertilised hermaphrodites. There is an ovotestis discharging into a common hermaphroditic duct, and separate sperm ducts and oviducts. Various adaptations promote copulation, and self-fertilisation occurs in very few species. As in all freshwater gastropods and most terrestrial invertebrates, the free-living larva has been suppressed and there is direct development within large yolky eggs.

10.10 Conclusion

Gastropods show greater diversity than any other group of invertebrates apart from arthropods. From a relatively unspecialised starting point, adaptive radiation (i.e. evolutionary divergence) has been most pronounced. At the same time, gastropods also provide very many clear examples of evolutionary convergence: in all groups similar changes have occurred from many different starting points — for example the changes associated with torsion in prosobranchs, the loss of shell combined with detorsion in opisthobranchs (in six different groups independently) and changes in shell shape in land snails. Even considering gastropods alone, the molluscan plan is shown to be most successful.

The next chapter considers the other two main groups of molluscs, which are very different from each other: Bivalvia and Cephalopoda illustrate extremes of divergence in the evolution of the molluscan body plan.

Chapter 11

Mollusca: Bivalvia and Cephalopoda

BIVALVIA

11.1 How is the molluscan body plan modified in bivalves?

In bivalves, the molluscan body is laterally compressed between the two halves of the dorsally hinged shell (Figure 11.1a). Characteristically the ctenidia are greatly enlarged, with elaborate ciliary tracts used for filter-feeding. The head and its associated sensory organs have been lost, as also has the radula, and feeding is assisted by ciliated 'labial palps'. The foot is lifted off the ground, becoming wedge-shaped; in sedentary forms glands in the foot secrete a 'byssus' of threads attaching the bivalve to the substratum. Muscles are strongly developed: a pair of adductors attached to the inner surfaces of the shell hold the two halves together, working against the elastic recoil of the hinge ligament, and anterior and posterior retractor muscles hold the byssus in place. Inhalant and exhalant siphons determine the flow of water in the mantle cavity.

Figure 11.1b,c,d indicates the structure of bivalves, with layers progressively removed.

11.2 What is the range of bivalves?

They are very common animals, with about 8000 known species. All are aquatic and most are marine, perhaps because the very large permeable gill surface presents a considerable osmotic problem for a freshwater bivalve. The common mussel *Mytilus* extends into

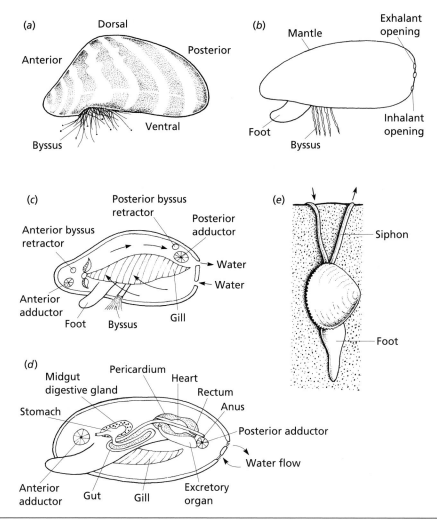

The structure of bivalves: (a) external view of zebra mussel (20 mm long); (b) diagram of mussel (*Mytilus*) with left shell valve removed, showing mantle; (c) diagram of mussel with left shell valve and mantle removed, to show water currents; (d) diagram of mussel showing visceral organs (nervous system omitted for clarity); (e) *Scrobicularia* burrowing, showing the long siphons.

estuaries, but few species live in fully fresh water. In the absence of a powerful excretory organ, a common adaptation is dilution of the internal body fluid and thus reduction of the osmotic gradient across the gills. The most dilute animal known is the freshwater clam *Anodonta*, with an internal fluid concentration one-tenth of that of a freshwater crayfish.

In size, bivalves range from a few millimetres to the giant clam *Tridacna* which is over a metre long. Here the exposed mantle edge contains symbiotic algae. Like those of the corals among which the clam lives, the algae may assist precipitation of calcium carbonate (needed to make the large shell) by removing carbon dioxide and reducing acidity.

The activity range is considerable also. Most bivalves are sedentary, i.e. buried or lying on a substrate but capable of limited movement. Some are sessile, i.e. entirely incapable of locomotion, such as the oysters, which are cemented to hard surfaces by one shell valve. Young stages are motile, i.e. capable of moving from place to place: planktonic larvae are important agents of dispersal, and even after metamorphosis young bivalves may move by extending the foot, attaching the cup-like sucker at its tip to the substratum and pulling the rest of the body over the foot. A similar technique is used by burrowing bivalves, such as razor shells (which burrow remarkably rapidly) and clams: the foot penetrates the sand, swells at the tip and takes hold, then the longitudinal muscles of the foot contract. In burrowing forms the siphons become elongated and reach up to the sand surface (Figure 11.1e). Other bivalves may bore through hard substances, using the anterior edges of the shell valves aided by chemical secretions. The ship worm *Teredo* bores through wood, and the chalk-borer *Pholas* is credited with having excavated the English Channel. Some bivalves even swim: for example the scallop *Pecten*, which has eyes all round the mantle edge, can take off and swim by clapping its shell valves together, expelling a jet of water.

11.3   How do bivalves feed?

Ciliary feeding by means of the ctenidia is characteristic, but probably the earliest bivalves fed like the **protobranchs**, deposit feeders where food is collected by the palp tentacles and sorted by ciliary tracts on the labial palps (e.g. *Nucula*, Figure 11.2a). A few protobranchs such as *Solemya* rely for food on chemosynthetic bacteria in the gills, with reduction or loss of the labial palps and part of the gut. The vast majority of bivalves, however, are 'lamellibranch': the name denotes not a division of the class but a condition of gill enlargement and elaboration, achieved many times separately within a number of families. The free edge of the lamella of each ctenidium is turned up and attached to the mantle or to the foot so that it becomes a fourfold compacted structure (Figure 11.2b). The gill surfaces become pleated, making grooves bearing two main ciliary tracts at right angles: the lateral cilia produce a water current and the frontal cilia trap food particles (aided by the long laterofrontal cilia) and beat them (in mucus secreted by the gill) to a ciliated anteroposterior groove, either at the top or at the free edge (Figure 11.2b). The food is then conveyed forward to the labial palps, where large particles drop down grooves and are rejected while small particles pass on to the mouth.

11.3.1   Digestion
Digestion is largely intracellular. The stomach is modified for the continuous arrival of small food particles bound together in mucous

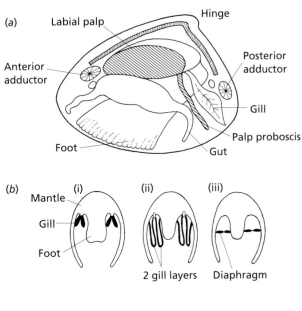

The functions of gills in bivalves: (a) *Nucula*, a protobranch, showing the large feeding palps and the small respiratory gills; (b) diagrams of bivalves in vertical section, showing different degrees of gill elaboration, (i) primitive condition, (ii) lamellibranch condition and (iii) septibranch; (c) transverse section of part of a loosely compacted double-layered gill (e.g. *Mytilus*).

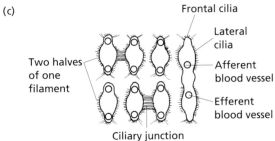

strings. A 'style sac' contains the 'crystalline style', a rod made of compacted mucus containing amylase and cellulase enzymes. As the style rotates by ciliary action against a chitinous 'gastric shield', the tip is continually worn away, with release of a small quantity of enzymes. The rotation also mixes the stomach contents and pulls in more mucous strings of food.

11.3.2   Septibranchs

These are a small group of deep-sea carnivorous bivalves that have abandoned ciliary feeding. The ctenidium has been lost and replaced by a muscular pumping diaphragm that draws in water carrying prey (Figure 11.2b). This is a bizarre reversal of the main direction of bivalve evolution, especially as it is combined with the crystalline style method of digestion, where there can be no extracellular protease. There are two deep-sea genera, *Cuspidaria* and *Poromya*.

11.4   What kinds of muscle are there in bivalves?

Bivalves need to maintain muscle tension at low metabolic cost, because the two halves of the shell need actively to be held together: when the adductor muscles relax, the elastic recoil of the hinge ligament opens the shell. Where there is a byssus, its retractors must similarly maintain tension. This need is met by a special form of muscle made of large 'paramyosin' filaments. It is slow to contract, inelastic and very isotonic, i.e. capable of maintaining tension at very different lengths (see Box 5.1). While it is very economical, the description of it as 'catch muscle' (implying that it can lock solid) is misleading because contraction does need to be maintained by occasional nerve impulses.

Bivalves also contain the normal invertebrate helical smooth muscle and occasionally fast-contracting striated muscle. The adductor muscle of the scallop *Pecten*, for example, can be seen to contain different regions: the more translucent is 75% striated muscle, used for quick twitches when swimming, and the more opaque region with long filaments is 56% paramyosin, used for holding the shell valves together.

SCAPHOPODA

11.5 How is the molluscan body plan modified in Scaphopoda?

Scaphopoda comprise a small (400 species) and very uniform group of marine molluscs encased in a one-piece tubular shell. Figure 11.3 shows *Dentalium*, a shallow-water genus much used for early embryological work. Scaphopods share some features with bivalves, but are sufficiently unlike other molluscs to be placed in a separate class. The mantle, together with the mantle cavity, surrounds the whole animal and secretes the tubular shell, leaving openings at both ends. Ctenidia have been lost; water flows in and out of the upper (posterior) end and respiratory exchange occurs in the mantle. The elongated foot emerges from the larger (anterior) aperture and is used for burrowing. The proboscis-like head bears a radula and clusters of contractile tentacles that collect small particles of food (Figure 11.3b).

CEPHALOPODA

These very remarkable molluscs have become jet-propelled predators in the sea, with the evolution of highly developed eyes, brains and behaviour. They are the only invertebrates to have filled the same ecological niche as fish, and they continue to fill it in competition with fish: squids too may be very numerous and form fast-swimming

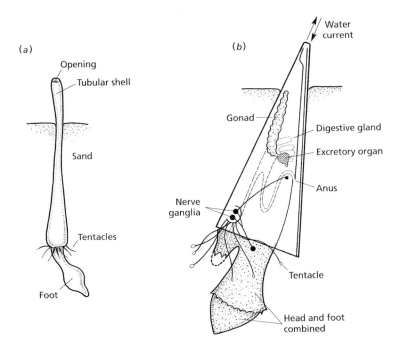

Fig. 11.3 The structure of scaphopods: (a) external view of *Dentalium*; (b) diagrammatic longitudinal section of *Dentalium*.

(a)

Opening

Tubular shell

Sand

Tentacles

Foot

(b)

Water current

Gonad

Digestive gland

Excretory organ

Anus

Nerve ganglia

Tentacle

Head and foot combined

shoals. Cephalopods also include the largest invertebrates: the body of a giant squid may be 15 metres long.

11.6 How is the molluscan body plan modified in Cephalopoda?

Whether the shell is retained, as in *Nautilus*, or internal and reduced or lost, as in coleoids, the animals are expanded along the dorsoventral axis and at the same time the ventral side becomes anterior (Figure 11.4a). The mantle cavity is ventral and opens anteriorly (behind the head). Part of this opening is filled by the funnel, which represents part of the foot, through which water is expelled in jet propulsion (see below). The ctenidia lack cilia, being ventilated by the stream of water drawn through the mantle cavity by contractions of the mantle wall. There is a closed blood circulation in place of an open haemocoel. The foot is represented by the funnel and by the muscular tentacles (arms) round the mouth. Prey is seized by the arms, masticated by beak-like jaws and transported further by the radula. Eyes are large and prominent.

11.7 What cephalopods are known?

Ammonites. These well-known and commonly found fossils are among a variety of extinct groups with external shells

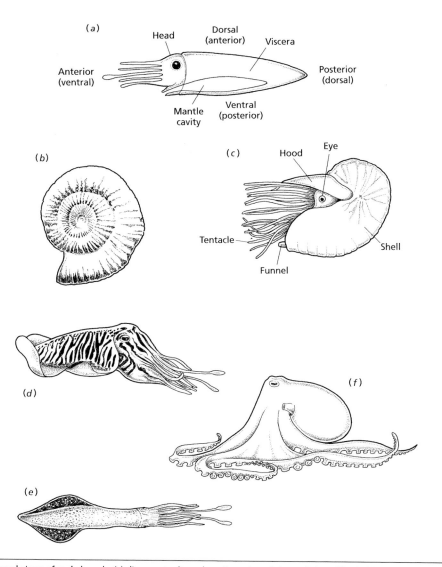

Fig. 11.4 External views of cephalopods: (a) diagram to show the orientation of the cephalopod body (the equivalents in other molluscs are given in parentheses); (b) a fossil ammonite; (c) *Nautilus*; (d) *Sepia*, a cuttlefish; (e) *Loligo*, a squid; (f) *Octopus*.

(Figure 11.4b). They became extinct at the end of the Cretaceous, as did the dinosaurs.

Nautiloids. The nautiloid group also have external shells but are not closely related to the ammonites. *Nautilus* (Figure 11.4c) is a present-day genus (with five species) which retains the external coiled shell, divided by septa into compartments. The animal lives in the open final compartment, with a thread of tissue, the 'siphuncle', extending back to the apex of the spiral. It is central in position, unlike the marginal siphuncle of ammonites.

Coleoids. These highly modified cephalopods (about 700 species) are without external shells (Figure 11.4d,e,f). They include:

Decapods, with ten arms, such as the cuttlefish *Sepia* (with an internal shell) and squids such as *Loligo* (with an internal horny rod in place of a shell).
Octopods, with eight arms, have no shell at all.

11.8 How is *Nautilus* able to survive?

Since, like many extinct cephalopods, it retains an external shell, *Nautilus* has been regarded as a deep-water relict, surviving by chance. Recent study reveals it to be a specialised modern animal with unusual adaptations (and that it is probably more closely related to the coleoids than to the ammonites).

a. **Buoyancy control**. Control of buoyancy is an important key to its success, as *Nautilus* can readily maintain or alter its position in the sea. (see Box 11.1).

Box 11.1 | Buoyancy

A small animal will float at the surface of water; it is said to be neutrally buoyant. Most animals, however, are heavier than water ('negatively buoyant') and tend to sink, especially in fresh water, which contains very little salt. An animal that must swim continuously in search of food or in order to ventilate its gills can correct the tendency to sink without extra energy cost, but this is rare. A large cuttlefish, for example, would need 10% more energy if it had no float. Most marine animals have adaptations enabling neutral buoyancy: jellyfish contain fewer heavy sulphate ions and more of the lighter chloride ions than the surrounding sea; siphonophores and various other animals have gas-filled floats; some squids have sacs storing ammonia; pelagic molluscs lose or reduce their shells; arthropods lighten their skeletons; and there are very many other examples.

Two groups of animals go further, having adjustable mechanisms giving them neutral buoyancy at different depths in the sea. These are those teleost fish which have gas-filled swimbladders, in which the pressure can be adjusted to match the hydrostatic pressure at different depths, and the cephalopod molluscs. The compartments of the shell, external in *Nautilus* and internal as the cuttlebone in cuttlefish such as *Sepia*, contain an adjustable mixture of gas and sea water. These cephalopods can maintain or alter their position in the sea with very little expenditure of energy; indeed, the ability to secrete gas into the shell may have been a crucial first step in the evolution of cephalopods. The shell is much thicker than in the ammonites, enabling *Nautilus* to live at depths down to 600 metres in tropical seas. Off the coral reefs of more temperate seas in the South Pacific, it is found at about 350 metres by day, where it is safe from turtles and other predators. At night it ascends to 150 metres, a level where there is more food. Buoyancy control is achieved by altering the proportions of gas and water in the shell compartments (see diagram a). When first formed, the chambers are full of sea water. The siphuncle takes up ions from the sea water by active transport (see Box 7.1), water follows by osmosis and enters the blood space within the siphuncular tube (see diagram b) to be swept down into the animal. In the chambers gas bubbles out of solution, replacing the water removed. The animal will then be more buoyant and float higher in the sea. When the siphuncle pumps ions in the opposite direction, from the blood to the chamber, water will follow and there will be less gas in the chambers; the animal will therefore be less buoyant and sink. In cuttlefish the chambers of the internal shell are lined by a membrane functioning very much as does the siphuncle in *Nautilus*, and buoyancy is regulated in much the same way.

Many fish also undergo daily vertical migration. They alter the pressure of gas in the swimbladder, rather than the proportions of gas and water; this process requires more energy and cannot give neutral buoyancy as quickly or as completely as is possible in *Nautilus*.

The control of buoyancy in *Nautilus*: (a) gas and water in the shell compartments; (b) the siphuncular tube enlarged, with an arrow indicating the direction of movements of ions and hence water, causing an increase in the proportion of gas.

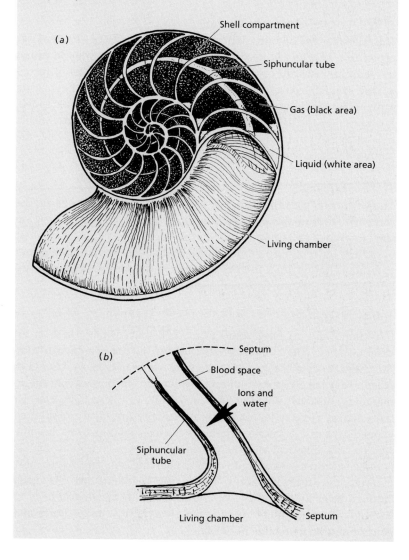

(a)

Shell compartment

Siphuncular tube

Gas (black area)

Liquid (white area)

Living chamber

(b)

Septum

Blood space

Ions and water

Siphuncular tube

Living chamber

Septum

b. **Metabolic economy**. A low metabolic rate is a striking attribute of *Nautilus*. It is an opportunist feeder, seizing crustaceans or small fish, and burns up its food unusually slowly. It can reduce this metabolic rate even further when starved and can continue to function at low oxygen tensions. *Nautilus* needs to live at low temperatures, below 20 °C. When very cold, the animal withdraws into the shell and remains motionless, building up an oxygen debt.

A slow pace of life is in general characteristic of *Nautilus*. Reproduction is by slow direct development in particularly large eggs. Further development is also slow, and *Nautilus* appears to live for longer than other cephalopods.

11.9   How have some cephalopods become so active?

11.9.1   Loss of the external shell
This allows great mobility: armour has been sacrificed to speed. Jet propulsion is achieved by strongly developed circular muscle round the mantle. Powerful contraction expels a fast jet of water, directed by the funnel: contraction of vertical muscle within the mantle wall allows water to enter and the mantle cavity is re-extended. In addition to the funnel jet there may be fins, which not only steady the cuttlefish and squids but by undulating waves provide the main method of slow movement.

11.9.2   Buoyancy
Cuttlefish are buoyant, using a mixture of gas and water in the compartments of the internal shell (see Box 11.1). Squids have no shell (merely a stiffening rod along the back) but may increase their buoyancy by retaining ammonia in special reservoirs (ammonium chloride is lighter than the sodium ions in sea water). Octopuses are not buoyant: they are bottom-living animals.

11.9.3   Defence
It is important to have some form of defence when there is no external shell. The sudden production of an ink screen can hide the animal. Coleoids can also become camouflaged by changing colour: they have 'chromatophores', which are elastic sacs containing pigments, usually blue, orange and yellow. These sacs can be flattened by contraction of a ring of radially arranged muscles, so expanding a disc of colour; when the muscles relax the sacs become smaller and spherical and the colour spots contract. Colour change is under nervous control and is very rapid. This is in contrast to most invertebrate chromatophores, where the pigments are dispersed through branching processes or concentrated in the central body of the cells; these pigment movements are typically under hormonal control and occur relatively slowly.

11.9.4   Feeding

Feeding is active and rapid. The arms are covered in strong suckers: decapods extend two long arms to seize their prey; octopuses jump onto a crab (or whatever the prey may be) entangling it in the interbrachial web. Toxins are secreted as the jaws bite.

11.9.5   Circulation of the blood

A closed circulation can be much more efficient, with capillary vessels between the arteries and veins. The blood pressure is relatively high (for a mollusc) and may be boosted by 'branchial hearts' at the bases of the gills.

11.9.6   Sense organs

The most remarkable evolutionary advance in cephalopods is the elaboration of the sense organs, nervous system and brain.

The suckers on the arms serve also as sensory receptors, the statocysts beside the brain detect gravity and changes in angular acceleration, and changes in water displacement are detected by a lateral line system, a remarkable parallel with fish. Rows of hairs parallel to the long axis of the visceral hump are continued over the head and along the arms: these receptors connect to sensory nerves, where the frequency of nerve impulses depends on the frequency and intensity of water movements. The most striking resemblance to vertebrates is shown by the eyes (Figure 11.5a). Apart from the external origins of the optic nerve, this could be the eye of a fish.

11.9.7   The brain

The cephalopod brain is the largest and most complex among invertebrates. Correlated with the sophisticated sensory equipment and complicated behaviour, the basic molluscan brain (a ring of nerve around the oesophagus) has become elaborated to form an organ that can be compared with that of vertebrates: indeed the brain to body weight ratio in a coleoid is greater than that of many fish and reptiles. The sea slug has some 20 000 neurons (nerve cells) divided into ten groups: the octopus brain contains an estimated 520 million, arranged in about a dozen separate lobes (Figure 11.5b,c), each with particular functions. Lobes concerned with motor output are ventral to the oesophagus; dorsal to it are lobes processing the various kinds of sensory input (touch, sight, etc.). The highest centres (absent in *Nautilus*) concern learning and long-term memory.

11.9.8   The nerves

The nervous system is correspondingly well developed. In *Loligo* the stellate ganglion (about 120 000 neurons) gives off branches all over the mantle, producing simultaneous muscle contraction. The giant nerve fibres, as much as a millimetre in diameter, conduct extremely quickly. Experiments on these nerves were the source of much of our

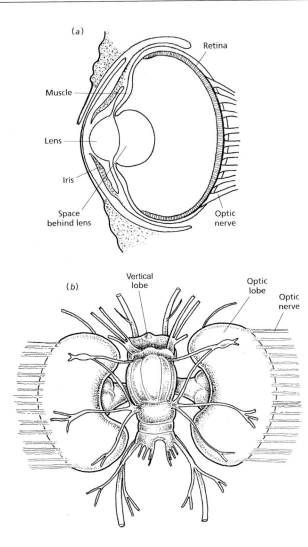

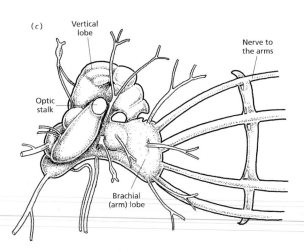

Fig. 11.5 (a) The eye of a cephalopod; (b) the brain of an octopus in dorsal view. (c) The brain of an octopus in lateral view.

knowledge about nerve action in general (see Box 11.2). With this equipment, the great learning power of coleoids is no surprise.

11.9.9 Reproduction

Reproduction has features unexpected in animals so advanced. The eggs may be brooded, as in the octopus, but the young never meet their parents. Development is direct in a large yolky egg (molluscs so mobile do not need larvae for dispersal). The behaviour of a newly hatched cephalopod is very stereotyped, and only later shows the flexibility that the brain allows. They live for one or two years, have a reproductive burst and then die.

Box 11.2 | Nerves and brains

Neurons (or neurones)

Most animals are coordinated by nervous systems. These consist of separate neurons that transmit information by conducting nerve impulses. Neurons can conduct impulses over long distances owing to their characteristic thread-like shape: typically there is a long fibre, the axon, leading from the nucleated cell body, which also bears shorter processes, the dendrites (see diagram). Ganglia are groups of neuron cell bodies; they include interneurons, which may or may not have axons but make connections between other neurons.

Synapses

These are the gaps at the ends of axons and dendrites. They conduct in one direction only, providing information to other nerve cell bodies or to other organs such as muscles. Transmission across neuromuscular junctions is usually chemical, often by means of acetylcholine, and after transmission the chemical is rapidly destroyed. Synapses within the nervous system may use chemical or electrical transmission; they may receive input from more than one neuron and may be either excitatory or inhibitory in effect.

Brains

Most animals have a coordinating nerve centre or brain, usually near the front end, to which 'afferent' nerves carry impulses from the sense organs and from which 'efferent' nerves carry impulses to muscles, glands or other 'effector' organs. Different animals have elaborated their brains to different extents according to their requirements.

The structure of neurons in (a) a vertebrate and (b) an invertebrate. The plan of the nervous system in (c) an acoelomate worm and (d) an annelid.

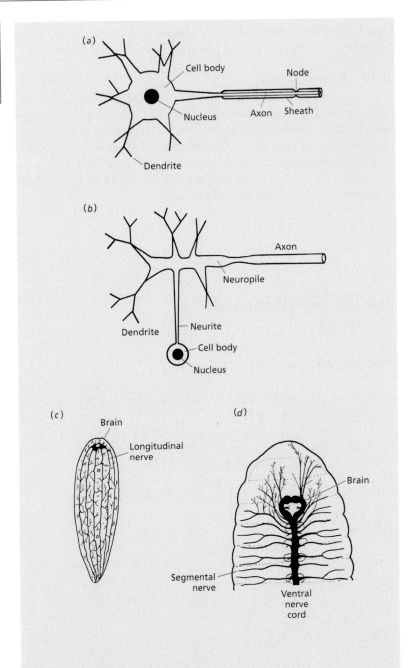

Neurosecretion

Many neurons are specialised for chemical secretion. For example, while vertebrate hormones are usually secreted into the blood by ductless (endocrine) glands, often under nerve control, invertebrates usually secrete hormones directly from nerve cells into the blood or tissue fluid. Neurosecretory cells have large nuclei and axons with terminal swellings beside fluid spaces. Moulting in arthropods (see Chapter 12), provides examples.

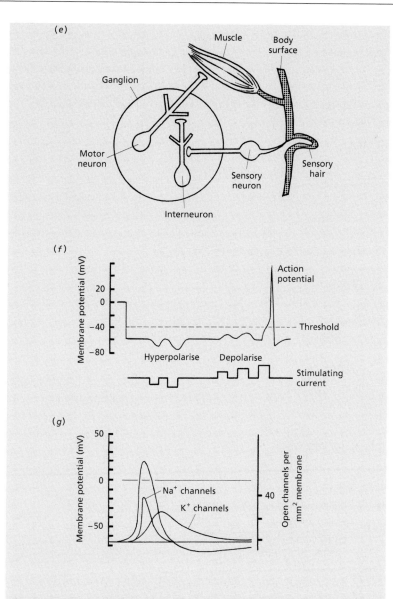

(e)

(e) The relations of a sensory neuron, an interneuron and a motor neuron; (f) the changes in membrane potential when a nerve is stimulated; (g) the passage of an action potential.

Invertebrate nervous systems

Sponges have no nervous systems. Cnidaria and echinoderms have simple nervous systems without brains. Most unsegmented worms have bilateral symmetry with some anterior sense organs and a brain that is little more than a sensory/motor link. Cell bodies and axons usually form about two main lateral or ventral nerve cords.

Annelids

Annelids have anterior ganglia above and below the oesophagus, linked by nerve loops called commissures. The supraoesophageal ganglion may be called the brain; it provides simple coordination between sensory input and motor output, and in some annelids (including earthworms) is not even needed to initiate movement. The suboesophageal ganglion is the first of a chain of segmental ganglia along the double ventral nerve cord (see diagram). Behaviour is largely based on segmental reflexes. Giant fibres with rapid conduction are important in burrowing worms and in tube-living polychaetes. Leeches have more specialised nervous systems, with the segmental ganglia condensed together and large ganglia under the suckers.

Molluscs

No generalisation can be made. Gastropods typically have an anterior 'cerebral' ganglion with very few controlling functions and about three other main ganglia, serving the viscera, the mantle and the foot. As well as nerve fibres connecting these ganglia there are longitudinal nerve cords, often twisted into a figure of eight by torsion. Sessile bivalves and active cephalopods represent extremes of brain development (see Chapter 11).

Arthropods

Arthropod nervous systems are built on the same plan as those of annelids (see diagram) but arthropod brains are usually highly developed and form the basis for most elaborate behaviour (see Chapters 12–15).

Chordates

Unlike other phyla, chordates have hollow dorsal nerve cords, though in invertebrate chordates these may occur only in the larval stages (see Chapter 18).

How nerves work

The basic facts are as follows:

1. Electric signalling along nerves is achieved by charged ions moving in and out of the axon (contrast the flow of electrons along a telegraph wire).
2. Passive ion movement is determined by chemical and electrical gradients.
3. Axon membranes are selectively permeable to sodium and potassium ions. There are separate sodium and potassium channels, each regulating the passive flow of the relevant ion.
4. Axon membranes, like the membranes of nearly all cells, have in addition a 'sodium pump': an active (energy-requiring) process that transports sodium ions outward, resulting in more sodium outside cells and more potassium inside them.

Resting nerve

The cell membrane is impermeable to the large internal protein molecules, which are predominantly negatively charged. Ions move freely in and out until equilibrium is reached. Due primarily to positively charged potassium ions moving out, the result is a potential difference of about 75 millivolts across the membrane, with the outside positively and the inside negatively charged. Sodium ions stay outside owing to the outward sodium pump. The membrane is said to be **polarised** (see diagram, f). Decrease in the resting potential is called **depolarisation**. Increase in the resting potential is called **hyperpolarisation**.

Nerve impulses

Impulses arise and are propagated when the resting nerve is stimulated, causing the sodium channels to open, so that sodium ions rush in. This not only depolarises the membrane but reverses the polarisation: the inside becomes positively charged (with a potential difference of about 55 millivolts). This is quickly reversed again by increased outflow of potassium, repolarising the active patch while the impulse is propagated along the nerve as an 'action potential' (see diagram, g).

This process is very economical, unlike muscle contraction. It requires only a very small amount of energy to restore the ion pools.

11.10   What has limited the evolution of cephalopods?

How is it that animals so highly evolved as active predators have not become even more successful? Part of the answer must be that they are limited by their molluscan constitution. For example, digestion is slow in the cuttlefish and the octopus, where the digestive gland is used both for enzyme secretion and for food absorption. Digestion of a meal takes 24 hours. In squids food is absorbed in the caecum also and the time is cut to four hours or less. The circulatory system, even with its modifications, may not supply oxygen to muscles very fast. Basic molluscan structure may be limiting in that flexible arms, with no joints as fixed reference points, cannot discriminate between shapes, however elaborate the brain: an octopus can have only limited powers of manipulation. Perhaps most importantly, the kidney remains that of a primitive mollusc, the blood pressure is not high enough for rapid filtration and the skin is too specialised to assist active ion transport or to keep water out. Cephalopods accordingly have negligible powers of osmoregulation and at the same time are too complex to tolerate changes in internal concentration. They are confined to the sea.

Perhaps we, descendants of the early vertebrates, should be relieved that there were limits to the performance of these extraordinary molluscs and that their degree of specialisation has prevented them from evolving further.

11.11　What are the evolutionary relationships of molluscs?

11.11.1　Relationships within the phylum

The Aculifera (molluscs with spicules, see Chapter 10) comprise the Aplacophora, with various features that might be primitive, and the Polyplacophora, which have shell plates very unlike the shells of other molluscan classes. The Monoplacophora may descend from the stem group of the Conchifera (molluscs with shells) but are not particularly close to the Aculifera. Molecular evidence suggests that the superficial resemblances between Scaphopoda and Bivalvia are due to convergence. Gastropoda, Bivalvia and Cephalopoda are extremely successful classes in very different ways: perhaps this diversity is made possible by the molluscan body plan, which although distinctive lacks constraining specialisation. With this diversity, and with no certain root for the phylum as a whole, perhaps it is not surprising that the verdict must be that the seven distinct classes are clearly related to each other but we do not yet know how.

11.11.2　Relationship with other phyla

Traditionally, the resemblance of molluscs to annelids in early development and their possession of a polychaete-like trochophore (Figure 10.3a) was seen as overwhelming evidence for a close relationship between these two protostome phyla, despite long-standing doubts about whether 'coeloms' in annelids and molluscs were homologous and whether molluscs were primitively segmented. More recently, emphasis shifted to the adult forms and it became commoner to derive molluscs directly from ancestral flatworms, creeping forms assumed to have evolved a protective shell dorsally and then to have needed a more complex structure. An attractive scenario, but one without supporting evidence. The earliest Cambrian fossil molluscs known were only 1−2 mm long, which suggests (but of course does not prove) that the stem molluscs were very small. Molecular evidence places the phylum in the large protostome group that includes both annelids and platyhelminths, without suggesting that they are particularly close to either of these phyla.

Chapter 12

Arthropoda: general

Arthropods are segmented animals with an outer cuticle, the exoskeleton, which is typically hardened and inflexible over much of the body but remains flexible at joints; the muscles are attached inside this cuticle. Like a molluscan shell it protects the soft parts, but, unlike a shell, the arthropod cuticle is built into the animal. With jointed limbs, fast-contracting striated muscle and a well-developed nervous system, rapid locomotion can be achieved.

The arthropod body plan has proved outstandingly successful. Arthropods constitute nearly nine-tenths of all known animal species, occurring in the sea, in fresh water, on land and in the air, in every conceivable ecological niche. Not only are they many and diverse as species (at least a million have been described) but also they are numerous as individuals, for example there are about 10^{21} copepod crustaceans in the oceans and some 200 million insects are said to exist for every human being on earth.

The outstanding success of arthropods is largely attributable to the nature of the cuticle. This is the primary theme of this introduction to the arthropods. The diversity of arthropods is described in the three following chapters, with emphasis on evolution from the sea to fresh water and land. Chapter 13 introduces the Crustacea, which are mainly aquatic. Chapter 14 includes Chelicerata, the marine horseshoe crabs and terrestrial arachnids, and also Myriapoda, all terrestrial. Chapter 15 discusses Insecta, which are nearly all terrestrial as adults, and often aerial.

Both the origins of arthropods and the relationships between them have been very controversial. Originally they were assumed to belong to a single phylum, derived from segmented annelid-like coelomate worms. Are they a single phylum or has this type of exoskeleton evolved several times separately? Should crustaceans, insects and chelicerates be placed in different phyla? After all, most of the characteristics defining an arthropod (such as jointed limbs, the need to moult, etc.) are necessary consequences of having a hard exoskeleton. Modern evidence increasingly confirms the original assumption of monophyly, but not the relationship to annelids. Chapter 20 combines fossil, molecular and morphological evidence to discuss arthropod relationships.

12.1   What defines an arthropod?

The exoskeleton covers the whole body. The body segments are often grouped into regions called 'tagmata' (singular 'tagma'), for example, head, thorax and abdomen. Most segments bear one pair of appendages: the most anterior segment may lack them, but primitively probably each of the more posterior segments bore paired appendages. In most groups, however, many appendages have been lost, as in many parasites and in the posterior region of insects. In general, arthropods with highly specialised appendages tend to have fewer of them.

The cuticle is moulted at intervals when the young animal increases in size; the process is controlled by hormones. The brain, sense organs and nervous system are extremely well developed: rapid and precisely controlled movement is characteristic of arthropods and complex behaviour has evolved. The main body cavity is a haemocoel containing blood, which enters the dorsally placed heart through holes (ostia). The coelom may be represented only in the gonads; it is segmental in origin but becomes much reduced during development, when the walls of the coelomic sacs break down. There are no locomotor cilia, but some sense organs are constructed from ciliary structures. As will be explained, respiratory and excretory organs differ greatly between aquatic and land forms. Typically the sexes are separate. The need for internal fertilisation is often associated with complex behaviour. Development is basically protostome (see Box 5.2) but much modified in most arthropods.

12.2   What are the key features of arthropod cuticle?

The frequent statement that 'to study cuticle is to study 90% of the properties of an insect' gives an emphasis that can be extended to all arthropods.

12.2.1   The cuticle

The cuticle or exoskeleton is made of chitin and protein, without any collagen. Chitin is a nitrogenous polysaccharide, related to mucus, from which perhaps it was derived. It is not in itself hard: microfibrils of chitin are embedded in protein and the cuticle acquires strength and rigidity when the protein molecule chains become cross-linked ('tanned') by phenols (quinones).

12.2.2   Cuticle is laid down by epidermal cells

The epidermal cells that lay down the cuticle are spread over the animal in a single-layered sheet. Each cell secretes its own overlying cuticle, giving precise spatial control.

12.2.3   Cuticle is laid down in layers

Low magnification reveals:

> **Epicuticle**: protein and wax, a protective outer surface (shed at moulting).
>
> **Exocuticle**: chitin and tanned protein, hard and often dark (shed at moulting).
>
> **Endocuticle**: ·chitin and soft protein (reabsorbed at moulting), with the epidermal cells underneath (Figure 12.1a).

The layers are crossed by ducts of dermal glands and by pore canals, probably conducting wax.

Electron microscopy reveals regular variations in the microstructure of the many layers constituting the exo- and endocuticle. Microfibrils of chitin and protein are orientated according to the mechanical demands on particular regions: most typically they change direction systematically through the thickness of the cuticle to form a helicoid (Figure 12.1b), an arrangement conferring strength and resistance to cracks.

12.2.4   Cuticle is extremely versatile

As well as supporting the animal and attaching muscles, parts of the cuticle need to be modified for special functions; for example, it is very hard and rigid over claws and biting mouthparts but thin and flexible at joints. It is capable of deformation where required, with elastic properties that enable recoil and storage of energy. It needs to protect the animal while at the same time admitting information from the environment. To achieve all this requires a formidably complex composite material.

12.2.5   Lipid

Lipid (wax) in the epicuticle forms the main barrier to water loss in terrestrial arthropods. Waterproofing with wax avoids the use of heavy materials. The wax layer is delicate: tanned proteins in the

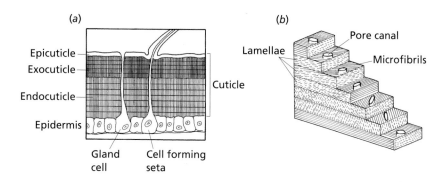

Fig. 12.1 Arthropod cuticle: (a) diagrammatic section; (b) diagrammatic section through a section of cuticle, showing the progressive change in orientation of the chitin–protein microfibrils in successive lamellae of the exocuticle, forming a helicoid structure.

cuticle below prevent deformation, but an insect cannot for instance go in and out of soil without abrading the epicuticle unless the wax has become protected by cement. The surface lipids serve also to deter predators and parasites, control temperature to some extent and deflect unwanted water. They may convey chemical signals to mates: the great variety of lipids in insect epicuticles suggests their use as pheromones (hormones acting outside the body). Epicuticular lipids are probably crucial to the survival of most arthropods.

12.2.6 Resilin

A uniquely elastic protein, resilin is capable of complete recovery of its shape after deformation. It consists of polypeptide coils linked at their sides by amino acids. Although the elastic properties of resilin (Figure 12.2) are especially vital to flying insects, resilin is also used by other arthropods, for example in the claws of scorpions, where there is no extensor muscle.

12.2.7 Modifications between groups

Different arthropod groups show characteristic modifications. Crustacea, mostly living in water where weight is not limiting, commonly have cuticles further hardened by calcium salts. In insects and arachnids, where water conservation is particularly important, the wax content of the epicuticle varies according to habitat, and can be altered by the animal in different conditions.

12.2.8 Moulting

Understanding of cuticle demands study of its formation in time as well as in space; study of moulting is revealing. To some extent arthropods can change in size and shape by making their exoskeleton

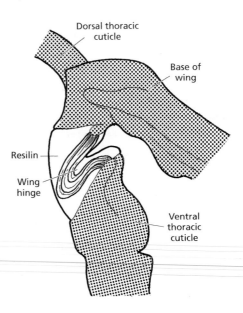

Fig. 12.2 Transverse section through the thoracic wall of *Schistocerca*, a locust, showing the pad of resilin at the wing hinge.

Dorsal thoracic cuticle

Base of wing

Resilin

Wing hinge

Ventral thoracic cuticle

more flexible, but once it is tanned, cuticle cannot be expanded. Moulting is dangerous, as the animal is briefly incapacitated and vulnerable, and wasteful of material. But larvae and their component structures such as mouthparts must increase in size, and this requires the shedding and replacement of the cuticle. The sequence of changes at moulting is shown in Figure 12.3. These changes culminate in ecdysis, the shedding of the old cuticle. The arthropod swallows some of the medium (water or air), pressure from inside splits the old exocuticle and the animal wriggles out. It quickly expands and tans the proteins in the outer layer of the new cuticle. The living tissues of the animal then grow and the cuticle is fully reconstructed.

The need to moult during growth provides an opportunity for 'metamorphosis', an abrupt change in form during the life cycle, important in many arthropods. Moulting and metamorphosis are

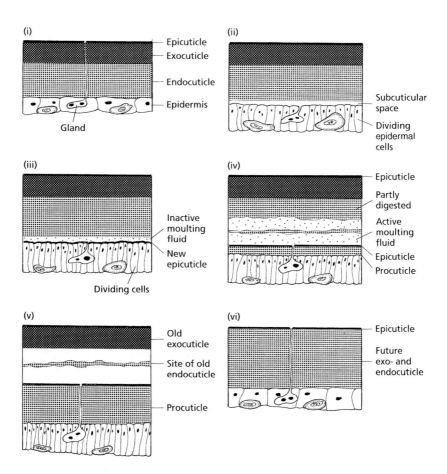

Fig. 12.3 Diagrammatic representation of the changes occurring in the cuticle during the moulting cycle: (i) mature cuticle; (ii) epidermis divides; (iii) new epicuticle layer produced; (iv) endocuticle digested; (v) moulting fluid resorbed; (vi) remains of old cuticle have been cast off.

under hormonal control. Figure 12.4 indicates the series of steps in insects and in crustaceans such as crabs, crayfish and prawns ('decapods'; see Chapter 13). The brain initiates neurosecretion of a hormone that is released into the blood from a structure near the brain (the corpus cardiacum in insects, the sinus gland within the eye stalk of crustaceans). In insects the brain hormone activates production of a moulting hormone precursor from the prothoracic gland, in decapod crustaceans the brain hormone inhibits hormone production by the Y organ. When the ecdysone precursor reaches the epidermis, ecdysone is formed and stimulates the cells to divide. Superimposed on these similarities are important

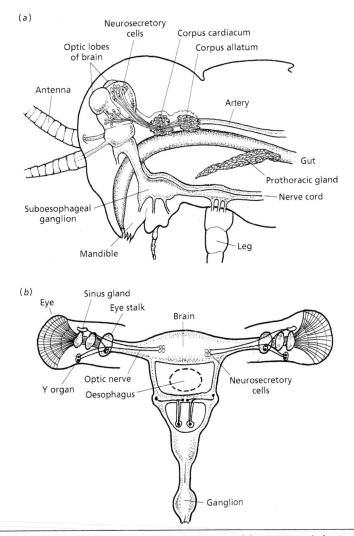

Fig. 12.4 Hormonal control of moulting: (a) diagrammatic longitudinal section of the anterior end of an insect, to show the brain and other sources of hormones controlling moulting; (b) diagrammatic dorsal view of the eye stalks and the anterior part of the nervous system in a decapod crustacean. Comparison of control of moulting in (c) an insect and (d) a decapod crustacean.

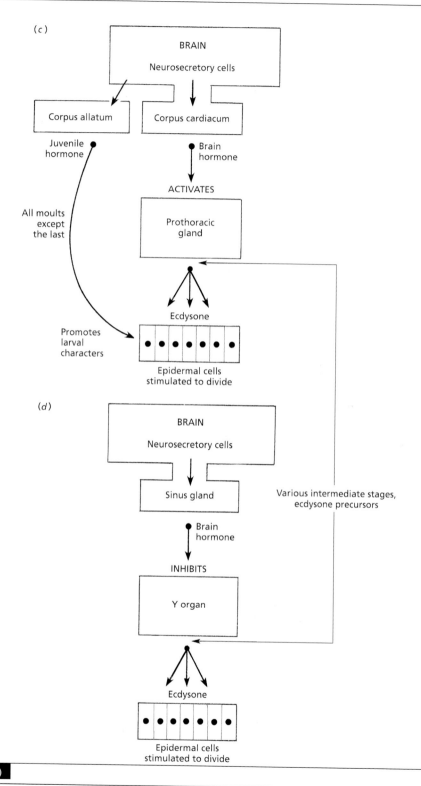

Fig. I2.4 (contd.)

differences: unlike many crustaceans, insects (except mayflies) never moult as adults but have a 'juvenile hormone' produced in a separate group of cells (the 'corpus allatum'). It diffuses to the epidermis and, at every moult except the last, promotes the appearance of larval form.

12.3   How are arthropod internal cavities organised?

The organisation of the soft parts of an arthropod is also largely dictated by the cuticle: for example, the absence of cilia is not surprising, since they could not work on the outside and are not needed on the inside. Cilia prevent cell division and therefore may direct or constrain development, a function redundant under a cuticle.

12.3.1   The haemocoel

With no requirement for a hydrostatic skeleton, the coelom is greatly reduced. It is represented only in the cavity of the gonads and, in some aquatic arthropods, in that of the excretory organs. The main body cavity is the haemocoel, divided into blood-filled spaces that bathe all the tissues (Figure 12.5). The blood is moved by muscular contraction. It enters the dorsally placed heart through the ostia and is pumped forward into vessels opening into the haemocoel; the heart beat is under nervous control in arthropods. This open system allows blood to be moved in bulk, which is important at moulting, providing turgor and maintaining the shape of the animal immediately after ecdysis. As well as dissolved food, hormones and other substances, the blood contains many cells, which may be stationary on surfaces or actively moving, such as the lurking phagocytes that constitute the immune system. In insects, where there is no separate digestive gland, other cells in the blood may store and process food.

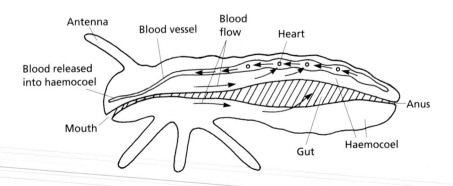

Fig. 12.5 Diagram of the open blood system in a generalised arthropod.

12.3.2 The gut

The gut (Figure 12.5) runs from end to end of an arthropod, with infolding of cuticle both at the mouth and at the anus. The midgut may receive secretions from digestive glands, and may be lined with permeable cuticle through which food is absorbed into the haemocoel. Gut structure varies according to the great variety of food and feeding methods.

12.4 What makes possible the great activity of arthropods?

12.4.1 The segmental plan

Segmentation is the subdivision of part of a developing embryo into sharply defined populations of cells, linearly arranged along the anterior/posterior axis. This process evolved many times separately, and segments are different kinds of units in different kinds of animal: for example, the cells primarily concerned with segmentation are mesodermal in annelids (see Chapter 9) but ectodermal, in the epidermis just beneath the cuticle, in arthropods. Segmental design can even apply to selected groups of cells in animals such as nematodes which have no overt morphological segmentation. What all these animals have in common is the involvement of similar groups of genes governing their formation.

Segmentation is associated functionally with locomotor efficiency. In arthropods the segmental plan combined with the hard exoskeleton allows rapid and precisely controlled movement. Unlike the muscular body wall of a worm, the muscles are discrete units controlled by segmental nerves. They move the segmental appendages: the most anterior appendages are usually sensory (e.g. antennae), the next few pairs are usually concerned with feeding (e.g. hard biting mouthparts) and more posteriorly there may be many pairs of walking or swimming limbs. Segmentation is frequently lost in the adults of sedentary animals, for example in many parasitic crustaceans.

12.4.2 The brain

This organ is greatly elaborated from the simple sensory/motor exchange found in the ganglia of many invertebrates. Behind it there is a nerve ring round the gut and a double ventral nerve cord with a ganglion in each segment receiving sensory innervation and giving off motor nerves, much as in annelids. Head development in arthropods is advanced, and the brain is an organ initiating and controlling behaviour. It is dorsal to the anterior end of the gut and divided into three lobes: the most dorsal receives sensory axons from the eyes, the middle lobe receives axons from the antennae (it is absent in chelicerates, which have no antennae) and the ventral lobe contains motor neurons giving off motor nerves (Figure 12.4).

Behind the brain there is a double ventral nerve with a pair of ganglia in each segment.

12.4.3   The neuromuscular system

All the muscle is striated. Like vertebrates, arthropods can move very fast with great efficiency, but they control their movements very differently. Arthropod muscle is unfamiliar to us in that it works inside a tube of exoskeleton, to which it is attached directly (often to infolded ridges of the cuticle called 'apodemes') or indirectly through tendons. Attached on either side of a hinge or a joint, muscles can build up tension and effect contraction with very little change in length, i.e. they are isometric (Figure 12.6a and Box 5.1).

Nerve control of muscle is far less centralised than in vertebrates: arthropod muscle response is finely adjusted at the nerve endings. A vertebrate limb muscle receives a large number of identical motor axons, each ending in a motor unit, and each muscle fibre is part of a motor unit. Contraction depends on the number of motor units stimulated and the frequency of stimulation, and any inhibition is also centralised. Arthropods have no motor units; instead a limb muscle receives few axons, which branch to more than one fibre and branch further to form 'twiglet' endings all over the muscle fibre (Figure 12.6b). The axons are not all alike: there are fast and slow fibres (the descriptions apply to the muscle response) and inhibitory ones. Vertebrate striated muscle contraction is 'all or none': if nerve stimulation reaches a threshold an action potential is generated and spreads all over the muscle, causing it to contract. In arthropods, muscle contraction is not all or none and muscle action potentials are rare: local 'junction potentials' occur under the twiglet nerve endings, causing a degree of contraction roughly proportional to the local electrical change. Muscle response is thus graded according to the local junction potential, which in turn depends

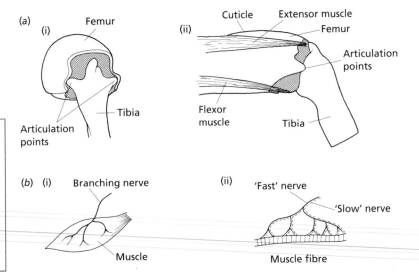

Fig. 12.6 (a) Arthropod (insect) leg joint, (i) to show articulation, (ii) diagram with cuticle of one side removed to show muscle attachments; (b) nerve control of muscle in arthropods: (i) nerve branching over the entire muscle, (ii) two different kinds of nerve with multiterminal twiglet endings over a single muscle fibre.

(a) (i) Femur — Articulation points — Tibia

(ii) Cuticle — Extensor muscle — Femur — Articulation points — Flexor muscle — Tibia

(b) (i) Branching nerve — Muscle

(ii) 'Fast' nerve — 'Slow' nerve — Muscle fibre

upon which axon has been stimulated, and whether or not it is inhibited at the site.

12.4.4   Sense organs: sensilla

An animal encased in armour needs structures sensitive to environmental events. The exoskeleton is locally modified into 'sensilla', in the form of hairs, bristles or pits that can receive chemical or mechanical stimuli and set up electrical changes in associated sensory neurons. Receptors detecting mechanical change are found at the base of appendages and elsewhere (Figure 12.7a,b).

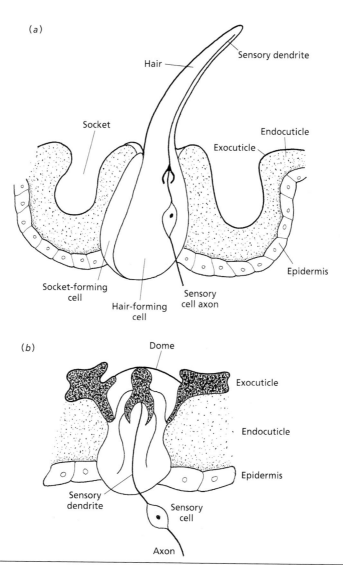

Fig. 12.7   Arthropod sense organs: (a) insect mechanoreceptor with sensory hair; (b) insect 'campaniform sensillum', a dome-topped structure detecting distortions in the nearby cuticle; (c) diagram of a compound eye; (d) diagrammatic longitudinal section of a single ommatidium.

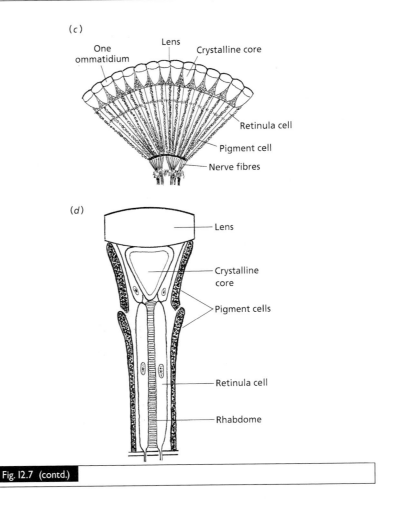

Fig. 12.7 (contd.)

Chemoreceptors resemble them but the sensory nerve extends to the tip of the hair: they are concentrated anteriorly, usually on the antennae, which may also contain sound receptors. In addition to these 'exteroceptors', there are at the joints 'proprioceptors' giving the animal precise information about the localisation and extent of its own movements. This allows a degree of control quite impossible for a worm, which cannot know exactly how its body bends.

12.4.5 The eyes

Compound eyes are characteristic of arthropods and occur in all groups. They are composed of a large number (as many as 4000 in some insects) of light-receiving units called ommatidia (Figure 12.7c). Each consists of a cylinder of elongated 'retinula' cells, with a small external lens (Figure 12.7d). The refractive index in the cylinder is graded so that the lens focuses light to the central axis or 'rhabdome', which is the much folded inner surface of the retinula cells. Here is situated the visual pigment, which responds to light by depolarisation (see Box 11.2) and which stimulates neurons to convey information to the brain. The ommatidia are separated by pigment, movable so that more or less light can be admitted.

A compound eye suggests a mosaic, with each ommatidium sensitive to a particular spot on the whole visual field. Most importantly, however, each ommatidium receives some of the light falling on its neighbours, so that the visual fields of neighbouring ommatidia overlap. Such a structure is extremely sensitive to movement, over a large field of vision, and this is the great advantage of a compound eye. It also provides magnification, and owing to the short light path may readily detect short wave lengths, for example ultraviolet light. Compared to the vertebrate (or cephalopod) eye it has, however, poor resolution and image formation.

The above generalised discussion indicates that most of the characteristic features of arthropods relate to their exoskeleton, and begins to explain the foundation for their diversity, both between and within the main groups.

12.5   What are the closest relations of arthropods?

The present-day arthropods are Crustacea, Chelicerata, Myriapoda and Insecta, and these will be introduced in the following chapters. Fossils, in particular one large fossil group, the trilobites, are the outstanding omission from this account (see Chapter 2).

Two small groups once included as 'arthropod-like', or 'proto-arthropods', are Pentastomida, now recognised as modified parasitic crustaceans (Chapter 13) and Pycnogonida, now included as chelicerates (Chapter 14).

Molecular and morphological evidence combines in identifying the two groups of non-arthropods closest to the arthropod ancestors, the Tardigrada and Onychophora.

12.5.1   Tardigrada

Despite some similarities such as the moulted cuticle, the tardigrades are not arthropods but constitute a separate phylum. These attractive little animals (Figure 12.8a) are the koalas of the invertebrate world. Some 700 species are known, only 0.1–1.2 mm long, living in water films on mosses or other plants in damp places or between particles in freshwater (less commonly marine) sediments. They feed mainly on bacteria. They have seven to eight segments bearing short lobe-like legs consisting mainly of claws, and a ventral nerve cord with segmental ganglia. Muscles radiate from the leg base and there is no circular muscle. These small animals have no true coelomic cavities, no blood vessels and no segmental excretory organs; structures resembling Malpighian tubules regulate the ion content of the body fluid and open into the gut. In dry conditions tardigrades survive by forming 'tuns', barrel-shaped resistant resting stages with greatly reduced metabolism; these may also be dispersal agents.

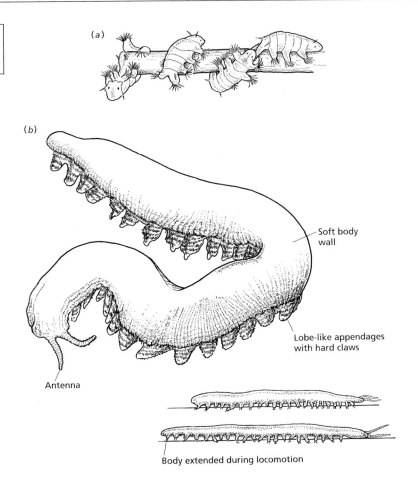

Fig. 12.8 Non-arthropods related to the phylum: (a) tardigrades; (b) *Peripatus*, an onychophoran.

(a)

(b)

Soft body wall

Lobe-like appendages with hard claws

Antenna

Body extended during locomotion

Tardigrades possibly evolved by miniaturisation, perhaps through early maturation of a small ancestral larva, or the small size may be primitive. Their ancestry is uncertain, but they appear to be close to the lineage of arthropods, within the group containing nematodes and some other animals with hard moulted cuticles.

12.5.2 Onychophora

Some 110 species of onychophorans are known, originally described as constituting a single genus, the 'velvet worm' *Peripatus* (body length 14–150 mm). These worm-like animals are considered to be the closest living relations of the arthropods. Living under damp logs, mostly in the countries of the southern hemisphere once united as Gondwanaland, they can vigorously squirt out mucus and look like slugs with legs: the soft body has no external signs of segmentation but is covered in soft (velvety) chitin, frequently moulted, with tubercles bearing iridescent scales. The body is, however, segmented, though without segmental nerve ganglia. There is only one pair of head appendages, the antennae, not homologous with those of arthropods. There are jaws round the ventral mouth and oral slime papillae, followed by 13 to 43 pairs of walking legs, which are unjointed lobes terminating in claws (Figure 12.8b). This combination of characters led to *Peripatus* being hailed as a 'missing link' between annelids and arthropods (when arthropods were thought to have arisen from annelid-like ancestors). Closer examination shows that *Peripatus* is a specialised modern animal adapted to a particular way of life: the body wall is lined by muscles working against hydrostatic pressure in the body cavity, as in many worms, making it very compressible. The animal is able to squeeze under stones and logs and through openings as small as one-ninth of its resting diameter. Yet the jaws and claws of these predatory carnivores have chitin, hardened by tanning of the associated protein, much as in arthropods. The body cavity is, as in arthropods, a haemocoel, the coelom being represented only in the cavities of the excretory and reproductive organs. Worm-like ciliated segmental tubules carry out excretion, yet respiration is achieved by arthropod-like tracheal tufts scattered all over the body. The spiracles cannot be closed and the cuticle is very permeable for a terrestrial animal, leading to great loss of water. Water absorption by the lobes between the legs is probably very important in enabling these animals to live on land, and they are confined to moist habitats. Reproduction in dry country is another problem. Remarkably, many Australian species are viviparous, with developing eggs not only retained for about six months but also supplied with food in the mother's body. Species in South America lay yolky eggs; there may be as many as 60 at a time.

The discovery of fossil marine 'lobopods' suggests the ancestry of this ancient and isolated phylum, separate from arthropods but close to their origin. Arthropod relationships are considered in Chapter 20, but first the four main groups of present-day arthropods will be discussed.

Chapter 13

Crustacea

Crustacea include crabs, lobsters, crayfish, woodlice, barnacles and many kinds of shrimp (Figure 13.1). They are primitively and predominantly marine: when some 38 000 species were known, 84% lived in the sea, 13% lived in fresh water and only 3% were terrestrial (there are now over 50 000 known, mainly due to discovery of more species in the sea). They are very successful arthropods, being extremely numerous as individuals and also diverse: in length alone, crustacea range from less than a millimetre to 4 metres (in some spider crabs). There are about 50 orders of crustaceans; the following account can only introduce a few of the forms in this great variety. Different lines of crustacean evolution (colonisation of fresh water and the land, and parasitism) are then discussed, and crustacean larvae are introduced.

13.1   What is distinctive about crustaceans?

How can they be recognised among other animals with the general arthropod characters? The main diagnostic character is that crustaceans have two pairs of antennae; if you find an obvious arthropod, count its antennae. There may be a first head segment with no appendages; the next two both bear antennae.

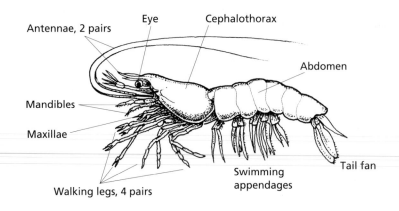

Fig. 13.1 A generalised malacostracan, based on *Astacus*.

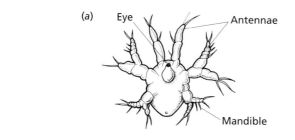

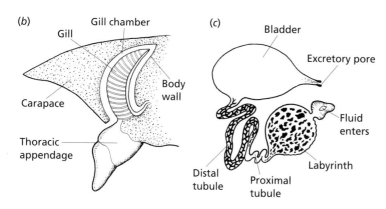

Fig. 13.2 Some crustacean features: (a) nauplius larva, ventral view; (b) a crab's gill; (c) the excretory organ of *Astacus*, a crayfish.

Typically but not always, the limbs are biramous (two-branched) and usually the head appendages include two pairs of maxillae.

Typically the life cycle includes a 'nauplius' stage, either as the earliest free-swimming larva with three pairs of appendages (Figure 13.2a), or as a stage contained within the developing egg. If there is a nauplius larva, the arthropod is a crustacean, but absence of the nauplius may be a secondary loss, especially in fresh water or on land.

Many characteristic crustacean features are associated with the aquatic habitat — for example, the hardening of the cuticle by calcium salts as well as by tanned proteins associated with the chitin — and are not found in land-living crustaceans such as woodlice. In aquatic crustaceans gaseous exchange for respiration occurs over the whole body surface (in small animals) or over the gills, which are usually modified basal parts of appendages, covered in soft permeable cuticle (Figure 13.2b). Oxygen is transported in the blood by the respiratory pigment haemocyanin. Excretion occurs partly by diffusion of ammonia over the gills or body wall and partly through mesodermal organs opening at the base of an anterior appendage (in crayfish these are called the 'green glands', Figure 13.2c).

13.2 What are the main kinds of crustacean?

Among a great deal of convergent evolution, one clear evolutionary trend can be recognised. Primitively, crustaceans were probably only a few millimetres in length, with a relatively large number of similar appendages. In present-day small filter-feeding forms all the appendages behind the antennae take part in feeding, locomotion and respiratory exchange, a condition believed to be primitive. From some such starting point, evolution has many times produced larger bottom-living forms that seek out and grasp their food in larger quantities. In many groups the head bears the segmental appendages shown in Figure 13.1: two pairs of antennae, the mandibles (hard biting structures made from the base of the third head appendage) and two pairs of maxillae, with perhaps some anterior trunk appendages assisting feeding as 'maxillipedes', and perhaps also 'chelae' (claws). In the more specialised Malacostraca, for example, (see below) the anterior end or 'cephalothorax' with the feeding appendages becomes enclosed in a shell or 'carapace', a cuticle-lined chamber protecting the gills, and the shortened trunk bears appendages specialised for walking or swimming. The extreme end point of these trends is represented by crabs, where the often thick carapace covers a body so short that the abdomen is tucked underneath and the four pairs of walking legs are so close together that the crab can only walk sideways.

Figures 13.3 to 13.6 give examples of the main groups of Crustacea:

Branchiopods, a very diverse group of small filter-feeding shrimps, many in fresh water, including the fairy shrimp *Chirocephalus* (Figure 13.3b), the brine shrimp *Artemia* and the water flea *Daphnia* (Figure 13.3c).

Copepods (Figure 13.4c), extremely common planktonic shrimps, mainly in the sea where they are a vital part of many food chains.

Ostracods (Figure 13.4d), a very separate and ancient group of crustaceans. Often less than a millimetre long, an ostracod is enclosed in a bivalve calcareous shell, even as a nauplius. The head bears unique appendages, the trunk has largely atrophied. Unlike water fleas, ostracods swim smoothly.

Cirripedes are unique and very much modified crustaceans: barnacles are all sessile, fastened to a substratum by their heads and kicking their food into their mouths with their legs. They include the extremely common sessile barnacle of rocky shores, *Balanus* (Figure 13.5) and the stalked ships' barnacle (goose barnacle), *Lepas*. The carapace has become a thick mantle over the trunk, the head attaches to the substratum and when the barnacle is under water the upper shell plates open and the legs shoot out, casting a food-collecting net. They are cross-fertilised hermaphrodites. Like all sessile adult animals, they depend for their dispersal on motile larval stages.

Malacostraca (Figure 13.6) constitute a large group of Crustacea with some 23 000 species. They are not particularly 'advanced' but include the biggest and best-known crustaceans.

Their anatomy is relatively constant: for example the body regions (head six, thorax eight, abdomen six segments) and the position of the genital openings (female sixth, male eighth abdominal segments). Malacostraca include some primitive shrimps with a seventh abdominal segment (e.g. *Nebalia*), the mantis shrimps (e.g. *Squilla*), the Australian freshwater forms with no carapace (e.g. *Anaspides*) and two main groups:

> **Peracarida**, with a ventral brood pouch in which the eggs develop, including shrimps such as *Mysis* (Figure 13.6a), **amphipods**, laterally compressed, e.g. sandhoppers such as *Gammarus* (Figure 13.6b), **isopods**, dorsoventrally compressed, e.g. *Ligia* at the top of the shore and the woodlice (Figure 13.6c) on land.

> **Eucarida**, with no brood pouch, the carapace being fused to the thoracic cuticle. These are the euphausids and the decapods. **Euphausids** include krill (Figure 13.6d), vital to the food chains in the southern oceans. Commercial fishing and melting ice (due to global warming) have seriously reduced populations over a wide area and to a worrying extent (in one area by 80% over the past 30 years). Penguins and whales are among the species at risk. Annual fishing limits are set, but hardly touch the problem. **Decapods** include the largest and most familiar crustaceans. The appendages of the first three thoracic segments have become maxillipeds assisting feeding, leaving five pairs of locomotor appendages (hence the name, 'ten legs').

13.3 How have crustaceans colonised fresh water and land?

13.3.1 Fresh water

Most branchiopods and the crayfish are freshwater animals, and most groups other than barnacles have a number of freshwater species, e.g. copepods, ostracods, amphipods, isopods, decapod shrimps and crabs.

They provide an interesting range of examples of the mechanisms (described in Box 7.1) by which freshwater animals resist or counter the tendency for water to enter and ions to leave their bodies. Crustacea can readily make the cuticle almost impermeable to water except over the gills. The body fluids may be considerably more dilute than in sea water. Dissolved salts depress the freezing point in direct proportion to their concentration. The depression of the freezing point is 1.8 °C in sea water and also in the internal body fluids of most marine crustaceans, but 0.8 °C in the crayfish *Astacus* and only 0.3 °C in the water flea *Daphnia*, as a small animal with a relatively larger surface area needs to be as dilute as possible. Crustaceans able to penetrate estuaries (e.g. the shore crab *Carcinus*) can tolerate considerable internal dilution by liberating stored amino acids from muscle cells into the blood. In fully fresh water, however, energy-consuming active ion uptake occurs at the gills (or over the whole body surface). This replaces ions leaving by diffusion, and excess water is removed by the excretory organ. A few freshwater crustaceans are also able to reabsorb ions in the excretory tubule, e.g. the amphipod *Gammarus pulex* and the crayfish *Astacus*, which unlike marine crustaceans have a distal portion of the tubule (Figure 13.2c) where ions are reabsorbed.

Most freshwater crustaceans have direct development. Rather than hatch as small fragile larvae, they brood large yolky eggs that hatch at a later stage, and the eggs may be very resistant to desiccation, hatching when supplied with water. The Chinese mitten crab *Eriocheir sinensis*, by contrast, migrates up and down estuaries

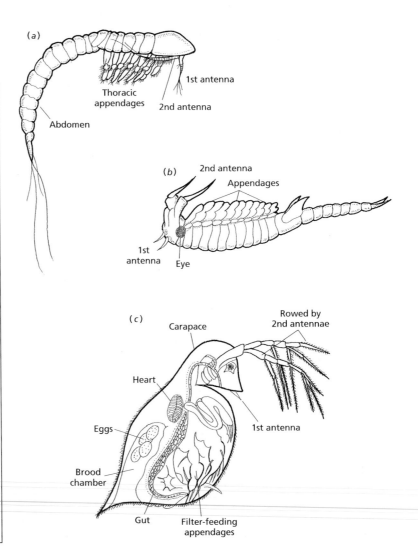

Fig. 13.3 The diversity of Crustacea. (a) *Hutchinsonia macracantha*: Cephalocarida are probably the most primitive living crustaceans; less than 3 mm in length, they live in marine mud. (b) *Chirocephalus grubei* (Branchiopoda, Anostraca), the freshwater fairy shrimp, swims with the dorsal side downward. Many leaf-like appendages, all used for feeding, swimming and respiration. On sale as pets, as 'sea monkeys'! (c) *Daphnia pulex* (Branchiopoda, Cladocera), the freshwater 'water flea'. (d) *Lasionectes entrichomas*: Remipedia are worm-like crustaceans less than 3 mm in length, first discovered in marine caves in 1981. Unspecialised in having many uniform trunk segments with many similar appendages, but probably this is secondary simplification. The brain structure is remarkably advanced. (e) *Derocheilocaris* sp., a mystacocaridan from the marine interstitial habitat, less than 1 mm in length. Only 10 species are known.

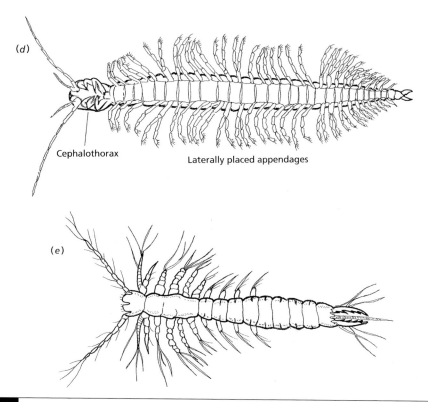

(d)

Cephalothorax   Laterally placed appendages

(e)

Fig. 13.3 (contd.)

during the life cycle, returning to the sea to breed, having therefore to withstand great osmotic change. It has the most impermeable crustacean cuticle known. Osmoregulation occurs at the gills, where ions subject to active transport may either be taken up (when the medium is more dilute than the crab) or extruded (when the medium is more concentrated). *Eriocheir* at the same time is an osmoconformer in that it can tolerate great changes in body fluid concentration. When the blood is diluted, amino acids released from muscle cells are broken down to proline and then to ammonium ions (NH_4^+) at the gills, where they are exchanged for incoming sodium ions (Na^+), saving energy. The urine is iso-osmotic with the blood: it seems remarkable that there is no ion reabsorption in the excretory organ, but calculations show that hypo-osmotic urine does not provide a significant energy saving in the fairly salty water where *Eriocheir* lives most of the time.

13.3.2 Land

Box 7.1 outlines the problems facing animals colonising land and the routes by which the transition was made. Few crustacean groups have fully terrestrial members. There are some land crabs, one family of amphipods and a profusion of isopods (there are at least

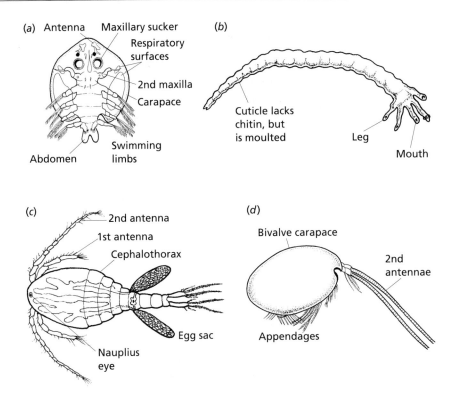

(a) Antenna Maxillary sucker
Respiratory surfaces
2nd maxilla
Carapace
Abdomen Swimming limbs

(b)
Cuticle lacks chitin, but is moulted
Leg
Mouth

(c)
2nd antenna
1st antenna
Cephalothorax
Nauplius eye
Egg sac

(d)
Bivalve carapace
2nd antennae
Appendages

Fig. 13.4 The diversity of Crustacea (continued). (a) *Argulus foliaceous* (Branchiura), ectoparasitic on fish: some 150 species, less than 30 mm in length; (b) *Cephalobaena tetrapoda* (Pentostoma) from a snake lung. Ninety species of these very much modified crustaceans are known, living as parasites in the nostrils or lungs of reptiles or other vertebrates. They were formerly placed in a separate phylum. (c) *Cyclops* sp. (Copepoda): there are some 8500 species and vast numbers of individuals of copepods in the marine plankton (and a few in fresh water), forming an important basis for many food chains. (d) An ostracod, about 1 mm in length, enclosed in a bivalve carapace. They are rowed along by their antennae, in the sea or in fresh water.

1000 species of woodlice). Habitats allowing semi-terrestrial existence were probably crucial to the transition: present-day semi-terrestrial crustaceans include the fiddler crabs (e.g. *Uca*), which spend much of their time in the air but retreat to water-filled burrows, and isopods such as *Ligia*, seeking out damp crevices high on rocky shores.

Terrestrial crustaceans have body fluids with high osmotic pressure, which at once suggests that they evolved not by way of fresh water but overland from the sea, perhaps over salt marsh or through mangrove swamps.

Land crabs are relatively large and impermeable invertebrates, not immediately liable to dry out. The stiffened gills do not entirely collapse in air and therefore allow some gaseous exchange, and part of the very vascular gill chamber functions as a lung. However, land crabs must return to the sea to breed, unlike amphipods and isopods, which retain their eggs in a brood pouch where they develop directly.

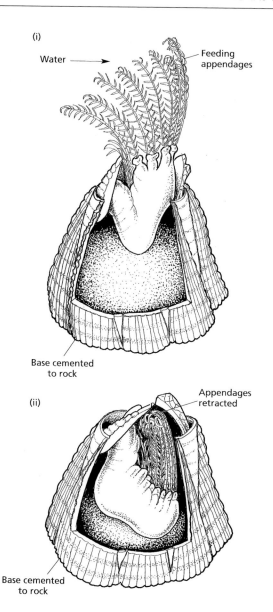

(i)

Water ⟶

Feeding appendages

Base cemented to rock

(ii)

Appendages retracted

Base cemented to rock

Fig. 13.5 The diversity of Crustacea (continued). *Balanus balanoides* (Cirripedia). Barnacles are unique, much modified sessile hermaphrodites, all marine: at high tide (i) the appendages are shot out and kick food into the mouth; when exposed to the air (ii), the appendages are retracted.

Amphipods of the family Talitridae are common above the strand line on sandy beaches: sandhoppers live in burrows above high tide mark by day and emerge at night when they feed on detritus. There is a series of species with progressively more enclosed respiratory organs, but terrestrial colonisation seems to depend mainly on tolerance of variation in body fluid composition and on behavioural adaptations.

Isopods have gills that are ventral flat plates at the posterior end of the body, covered and enclosed by the last appendages (uropods). Air-filled 'pseudotracheae' are probably crucial for breathing air with minimal water loss. Woodlice tolerate changes in internal concentration, and have the unusual ability

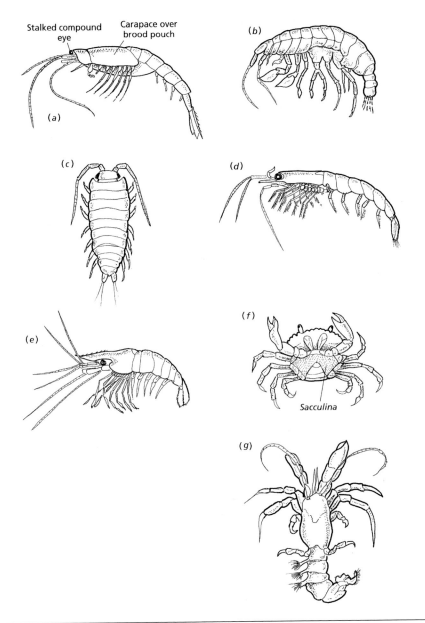

Stalked compound eye

Carapace over brood pouch

(a)

(b)

(c)

(d)

(e)

(f)

Sacculina

(g)

Fig. 13.6   The diversity of Crustacea (continued): some Malacostraca. (1) Peracarida, with brood pouch: (a) *Mysis relicta*, an opossum shrimp; (b) *Gammarus locusta*, a sandhopper (Amphipoda, laterally compressed); (c) *Armadillidium*, a woodlouse (Isopoda, dorsoventrally compressed). (2) Eucarida, with the carapace fused to the thoracic cuticle and no brood pouch; (d) *Euphausia superba*, shrimp-like, e.g. krill, the main basis of Antarctic food chains. Decapod eucaridans have five pairs of locomotor appendages, since the first three thoracic pairs assist in feeding; (e) *Leander squilla*, a prawn; (f) *Carcinus maenas*, the shore crab, ventral view, with the rhizocephalan parasite *Sacculina carcini*; (g) *Eupagurus bernhardus*, a hermit crab, showing the soft, reduced, asymmetrical abdomen.

to excrete by puffing out gaseous ammonia. Woodlice have become very successful terrestrial animals, but cannot live in hot dry air. They have to have behavioural adaptations that keep them in damp places.

13.4 What may limit the size of Crustacea?

Aquatic arthropods can become much larger than terrestrial ones: the exoskeleton is supported in water, and respiration by gills, unlike aerial respiration, does not demand small size. Accordingly many marine crustaceans, bottom-living, pelagic and even intertidal species, have been able to profit from the ecological and metabolic advantages of increase in size.

The largest examples of many groups are found in the Antarctic. There are no decapod crustaceans, but their place is taken by a very large number of amphipods and isopods. Amphipods in particular have been studied by scientists from the British Antarctic Survey, who investigated the finding that the largest polar species have bodies five times the length of amphipods in warmer waters. The solubility of oxygen in sea water is considerably increased at low temperature: it is 1.82 times greater at 0 °C than at 30 °C. Increased availability of oxygen, rather than a direct effect of low temperature, was postulated as the factor allowing size increase. The effects of low temperature and high oxygen content can be distinguished by comparing freshwater lakes at different altitudes: in Lake Titicaca (altitude 3809 metres, in the Andes) the oxygen content is reduced by high altitude, and also by relatively large quantities of dissolved salts. There the maximum length of amphipods is 22 millimetres. In Lake Baikal (in the lowlands of Russia, with no measurable salt content) the amphipods are 80–90 millimetres long, four times larger than in Lake Titicaca. From the temperature difference alone (12 °C in Lake Titicaca, 6 °C in Lake Baikal) only half this increase in size would be expected. This is strong evidence that oxygen availability rather than temperature is the limiting factor for maximum size in amphipods.

Amphipods do not have very efficient mechanisms of acquiring oxygen. It diffuses into the blood over the gills and is carried partly in solution in the blood and partly in combination with haemocyanin, but the circulatory system is restricted, with few lateral branches. With this system it is not surprising that availability of oxygen limits the size of amphipods (obviously this finding cannot be generalised to other crustaceans).

At the other end of the scale, the minimum size is not related either to oxygen supply or to temperature; it probably depends on the minimum possible size for an invertebrate egg.

13.5 What are the special features of parasitic crustaceans?

Parasitism has arisen many times separately in Crustacea. Ectoparasites, living mainly on the gills or the outer surface of fish, are recognisable crustaceans with certain appendages modified for attachment to the host (e.g. Branchiura, several separate orders of copepods and also Tentaculocarida, which are a whole newly discovered class of deep-sea crustaceans). Endoparasitism, however, demands radical reorganisation, since an unmodified exoskeleton would be a barrier between parasite and host. For example, the 90 species of Pentastomida (Figure 13.4b), parasites in the nostrils or lungs of reptiles, were formerly placed in a separate phylum. The cuticle lacks chitin but is moulted, and there are no respiratory, circulatory or excretory organs. *Sacculina* is an even more modified example. The body consists of an external sac of eggs under a crab abdomen (see Figure 13.6f) and a series of food-absorbing tubes inside the crab. It is revealed as a crustacean only by the nauplius larva, which has the 'horns' characteristic of cirripede nauplii (Figure 13.7a). The host is not killed, but the male is progressively more feminised at each moult as the parasite consumes the androgenic gland, the source of male hormones. Other Rhizocephala are parasites with little more than fungus-like rooting and food-absorbing systems inside the host's body. Metamorphosis in these animals is perhaps more drastic than in any other parasite, with no larval tissue being carried through to the adult.

Copepod endoparasites tend similarly to be very much reduced. Among them Monstrilloida, in a variety of invertebrate hosts, are unusual in being parasitic only as juveniles. The adults are free-living and recognisable as crustaceans, but lack mouthparts. Yet other crustaceans have division of labour between the sexes, only the female being parasitic.

In general, Crustacea require great specialisation (in the guise of 'degeneration'; see Box 6.1) to become successful endoparasites. Success is facilitated by the need to moult repeatedly during the life history, since a series of larvae can effect transfer between hosts.

13.6   What is the role of crustacean larvae?

The need to moult provides opportunities for changes in form, often striking enough to be called metamorphosis. Different forms in the life cycle between the egg and the sexually mature adult allow a division of labour: most typically crustacean larvae are feeding stages in a habitat different from that of the adult, avoiding competition with the adult, and serving also to disperse the animals. Most marine Crustacea hatch as planktonic larvae, in the surface waters which are rich in food and have currents separate from those lower in the sea. Freshwater Crustacea seldom have larvae and terrestrial ones never do (see above). Large yolky eggs contain and feed the developing young, often within a brood pouch.

Figure 13.7 shows some examples of crustacean larvae at different stages. The nauplius is most commonly the first larval stage. It has three pairs of appendages, the smallest number that can filter-feed. Even within a directly developing egg, a pause at a recognisable nauplius stage can often be detected.

Copepods hatch as a nauplius, followed by a metanauplius with four pairs of appendages, and then a series of stages progressively more like the adult. Metamorphosis is very gradual, with adults remaining in the larval habitat. They retain some larval characters, for example there is no carapace and only a simple 'nauplius' eye.

Cirripedes have a far-reaching metamorphosis. After the nauplius, distinctive with 'horns' (Figure 13.7a), and metanauplius there is a 'cypris' larva (Figure 13.7b) that does not feed but selects the adult habitat, seeking a rock where a protein 'message' has been left by an earlier barnacle occupant. In a sessile animal, larvae are particularly important for dispersal and habitat selection.

Malacostraca nearly all hatch beyond the nauplius stage. Freshwater malacostracans have direct development, marine ones often hatch as a zoea (Figure 13.7c) followed by a larva more like the adult. For example, the crab zoea is followed by a megalopa (Figure 13.7d), a small crab with the abdomen extended instead of being tucked underneath.

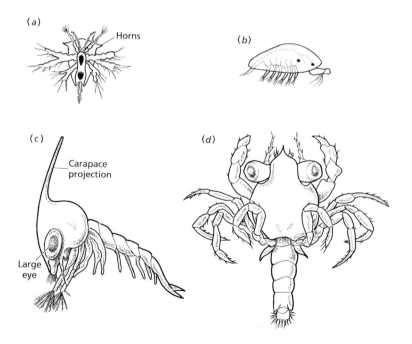

(a)

Horns

(b)

(c)

Carapace projection

Large eye

(d)

Fig. 13.7 Some crustacean larvae: (a) cirripede (barnacle) nauplius; (b) cirripede cypris, which does not feed but searches for a site and settles; (c) crab zoea, the first larval stage; (d) crab megalopa, a later larval stage, with abdomen still extended.

13.7 How are Crustacea related to each other?

Faced with so much diversity, there is bound to be controversy about theories based on morphology. While recent fossil finds and molecular evidence can help, many uncertainties remain: for a start, which present-day crustaceans are closest to the primitive condition? Probably the earliest crustaceans were very small and unspecialised. A recently found early Cambrian fossil, *Ercaia minuta*, was 2–4 mm in length and had 13 similar segments, all with similar appendages – but we cannot assume that it represents the primitive structure just because it is the earliest crustacean known. Unspecialised small present-day crustaceans such as Cephalocarida or Remipedia seem to be obvious candidates, but these groups may be secondarily simple (for example Remipedia have brains as elaborate as those of some Malacostraca). Concerning further crustacean evolution, also, uncertainty remains. Groups recognised in recent classifications as being monophyletic, using morphological and molecular evidence, are Branchiopoda, Cephalocarida, Remipedia and Malacostraca. Opinions differ about the other main groups: some workers combine Copepoda, Cirripedia, Mystacocarida, Branchiura and perhaps Pentostomida as 'Maxillopoda', with or without the inclusion of Ostracoda. Figures 13.3 to 13.6 illustrate all these groups, but cannot indicate the diversity which many of them represent.

Crustacea, in summary, are a very large and diverse group of essentially aquatic arthropods. The arthropod groups to be considered next have members far more thoroughly adapted to life on land.

Chapter 14

Chelicerata and Myriapoda

CHELICERATA

While Crustacea are aquatic arthropods with a few terrestrial members, Chelicerata (*c.* 75 000 species) are terrestrial arthropods with a few aquatic representatives. They demonstrate one version of the adaptations necessary for a terrestrial arthropod.

14.1   What are chelicerates?

They constitute a very distinctive group of arthropods whose segmental plan is different from others in that they have no head, only one pair of mouth parts (yet they are predators) and, unlike all other groups of arthropods, no antennae. The body is divided into an anterior prosoma and a posterior opisthosoma. The prosoma usually consists of eight segments. The first segment never bears appendages, the second bears paired chelicerae (feeding and grasping structures) and the third bears paired pedipalpi (sensory, prehensile or reproductive appendages). The fourth to seventh segments bear walking legs (i.e. four pairs) which have small pincers (chelae) at their ends. Chelicerates (Figures 14.1, 14.3) comprise:

> **Merostomata**, aquatic forms with five or six pairs of opisthosomal
> appendages modified as gills, consisting of:
> > **Eurypterida** (sea scorpions), all extinct.
> > **Xiphosura** (represented today by the horseshoe crab *Limulus*).
> **Pycnogonida** ('sea spiders'), superficially resembling spiders.
> **Arachnida**, all the terrestrial chelicerates, including:
> > **Scorpions**
> > **Pseudoscorpions**
> > **Solifugae** (sun spiders)
> > **Opiliones** (harvestmen)
> > **Acari** (mites and ticks)
> > **Araneae** (spiders).

14.2 Why is *Limulus* of special interest?

Comparisons can be made both with marine crustaceans and with the terrestrial chelicerates. *Limulus* (Figure 14.1a,b) is one of the three genera of marine chelicerates alive today. It burrows into the upper layers of sand in shallow seas, using the shield-shaped prosoma that protects the ventral appendages: these are the chelicerae and five pairs of walking legs (the front pair corresponds to the pedipalps of other chelicerates). At the posterior end of the gill-bearing opisthosoma there is a long spike, used for righting the animal and pushing.

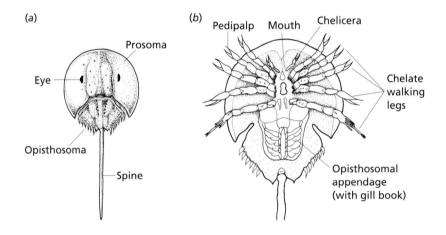

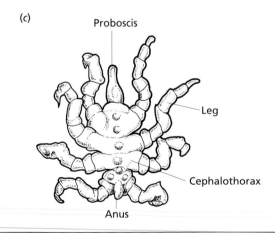

Fig. 14.1 Aquatic Chelicerata: (a) *Limulus polyphemus*, the horseshoe crab, dorsal view; (b) *Limulus* in ventral view, showing appendages: opisthosomal appendages bear gills, in a respiratory chamber; (c) *Nymphon rubrum*, a pycnogonid.

14.2.1 Feeding

Molluscs, worms or other animals are picked up by the chelate appendages and held against a ventral groove in the body. The food is chewed not by mandibles but by hardened 'gnathobases' of the posterior walking legs and passed forward to the mouth, where it is swallowed and then further crushed in a gizzard.

14.2.2 Respiration

Opisthosomal appendages are modified into 'gill books' (Figure 14.1b) that move to and fro, making a continual stream of water over the gills. The blood contains the respiratory pigment haemocyanin, which has a low oxygen affinity, able to function at extremes of environmental oxygen pressure. Blood enters the long heart through eight pairs of ostia with valves. The further circulation is complex, with many arteries opening into the haemocoel.

14.2.3 Excretion and osmoregulation

As in Crustacea and other aquatic arthropods, nitrogenous waste is removed partly as ammonia diffusing over the gills and partly through mesodermal glands at the base of a pair of appendages (here 'coxal glands' beside the back legs). *Limulus* can tolerate a wide range of salinities: the gills can take up ions when the medium is diluted and the coxal glands can produce urine hypo-osmotic to the blood. The assumption, based on the similarity of the cellular structure to that of crustaceans, is that the cortex forms an ultrafiltrate that is modified in the medulla and end sac.

14.2.4 Reproduction

The sexes are separate. Males and females aggregate in shallow water at high tide when the moon is full. Fertilisation is external: the male mounts the female and pours sperm over the thousands of eggs as she lays them. The eggs develop in the sand and hatch as 'trilobite' larvae which swim and burrow and, in a series of moults, undergo gradual metamorphosis.

14.2.5 The compound eyes

First studied early in the twentieth century, the relatively uncomplicated compound eyes of *Limulus* were the primary source of information about these arthropod structures, and remain the context in which reference to *Limulus* is most common. *Limulus'* compound eyes do not need to work very fast. They are used not to find food, nor to escape from predators, but to detect the movement of other horseshoe crabs across the field of vision; they are apparently used only to achieve the aggregation of mating males and females in shallow water at high tide. Accordingly, the structure and performance of this eye (Figure 14.2) is relatively simple and accessible to study.

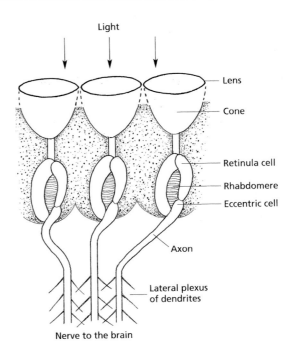

Fig. 14.2 Diagram showing three ommatidia of the compound eye of *Limulus*.

In summary, *Limulus* is a specialised modern animal, but none the less it gives us an aquatic 'baseline' for considering the Arachnida.

14.3 What are pycnogonids?

Pycnogonids found on our shores are small (1—10 mm long) arthropods crawling on sponges, hydroids or bryozoans. About 1000 species are known. Deep-sea forms may be up to 40 mm long in the deep sea, and much more in the Antarctic, where they are very common. Pycnogonids are called sea spiders because they typically have four pairs of very long legs, each borne on a large extension of the supporting segment (Figure 14.1c). Usually they have little else: the body typically consists only of a narrow prosoma with an anterior sucking proboscis (they are carnivores or detritus feeders). The anterior appendages are sufficiently like chelicerae and palps to diagnose them as chelicerates, but they are highly modified and of unknown relationship to Merostomata and Arachnida. The opisthosoma is remarkably reduced or absent, the digestive organs and the gonads extend into the legs and there are no special respiratory or excretory structures. The ova ripen in the legs of females and the eggs are carried by the males on modified appendages in front of the first pair of walking legs.

14.4 What are arachnids?

All terrestrial chelicerates are included, and the group is thought to be monophyletic. Ninety-five per cent of the species are spiders, mites and harvestmen. Typically arachnids are predators feeding on other arthropods. Rather than catching them by sustained rapid locomotion, they use poisons – neurotoxins that paralyse their prey. The poison may come from the chelicerae or pedipalps or from a posterior sting – arachnids may be dangerous at both ends. Digestion often occurs outside the body, with the posterior end of the oesophagus modified as a 'pumping stomach' to suck in the digested food.

> **Scorpions** (1000 species). Scorpions are large arachnids, up to 90 mm (one species, 210 mm) long with pedipalps bearing large characteristic chelae and with an extended segmented opisthosoma ending in a sting (Figure 14.3a). They are nocturnal hunters, burrowing or hiding under damp logs or in leaf litter. Most require humid conditions but some can live in deserts. They have a number of characters primitive among chelicerates, such as compound eyes and a nerve cord with

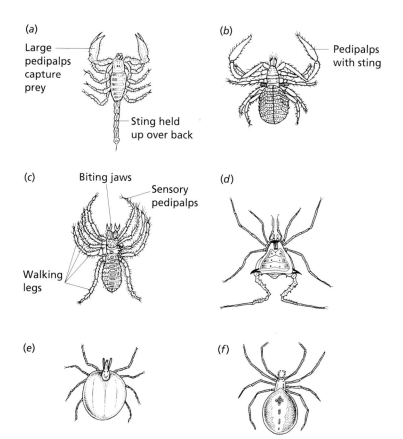

(a) Large pedipalps capture prey — Sting held up over back

(b) Pedipalps with sting

(c) Biting jaws — Sensory pedipalps — Walking legs

(d)

(e)

(f)

Fig. 14.3 Arachnida. (a) A scorpion. (b) A pseudoscorpion: there are about 2500 species, less than 8 mm in length, in crevices and leaf litter; their resemblance to scorpions is only superficial. (c) A solifugid: about 900 species, up to 7 mm in length; they are fast-running desert predators, with long sensory pedipalps held out in front and powerful biting pedipalps but no poison. (d) An opilionid: there are 4500 species of 'harvestmen', 5–20 mm in length, predators or scavengers in damp grass or forests. (e) A tick: there are 30 000 species of mites, ticks and other Acari, very numerous as individuals, less than 1 mm in length, free-living or ectoparasitic. (f) A spider.

separate segmental ganglia rather than a fused ganglionic mass. They resemble the fossil aquatic eurypterids ('sea scorpions') and are known from the Silurian, as fossil aquatic scorpions with gill books protected by opisthosomal plates. Modern scorpions are all terrestrial, with the gill books replaced by lung books and stout sensory setae replaced by long thin sensory hairs. Scorpions have unique paired sensory combs ('pectines') that may be chemosensory or may detect vibrations in the ground.

The main characteristics of **pseudoscorpions, sun spiders, harvestmen** and **mites and ticks** are indicated in Figure 14.3.

Spiders (32 000 species). Spiders are mostly 2–10 mm long, but tarantulas measure up to 90 mm. They are all carnivorous, feeding almost exclusively on insects. They are extremely successful terrestrial predators, largely on account of their production and use of silk (see below). The prosoma is convex dorsally, with marginal eyes (typically eight). The chelicerae seize the prey and paralyse it with poison, the pedipalps bear projections that manipulate food and may wrap it in silk (until it is ready for external digestion by enzymes poured out by the spider) and there are four pairs of walking legs (Figure 14.3f). A characteristic waist or 'pedicel' separates the prosoma from the opisthosoma, where soft parts of the body are housed and silk is manufactured and released.

14.5 How did arachnids colonise the land?

Land was probably colonised many times separately; the relationship between arachnid orders is entirely obscure. Scorpions in the Silurian are the earliest known terrestrial arachnids; spiders are found in the early Carboniferous. The blood of all arachnids has a high osmotic pressure and a composition similar to that of *Limulus*, with sodium and chloride ions predominating: this suggests marine ancestry and direct colonisation of land over the upper shore (see Box 7.1). Spiders are especially successful terrestrial arthropods. Insects had appeared in the mid Devonian, and the relationship between the two groups must have been established early, with increase in spider numbers following the radiation of insects. Today spiders are major invertebrate predators, numerous in all terrestrial ecosystems; for example a study of untreated grassland in late summer in Sussex, England gave an estimate of 2.25 million spiders per acre (about 5.55 million per hectare).

Arachnids are adapted to terrestrial life mainly by physiological changes and also by their behaviour: relatively large arachnids often burrow, and smaller ones are 'cryptozoic', hiding in crevices or under rotting logs. Physiological changes (as in other arthropods) primarily concern methods of resisting desiccation by modifications to the cuticle, the respiratory system and the excretory organs; sense organs and reproductive methods also need to be different in a terrestrial habitat.

14.5.1 The cuticle

Like that of insects, the arachnid cuticle has an outer layer containing wax, the epicuticle, which to varying extents resists water loss: some scorpions and ticks are as impermeable as insects.

14.5.2 Respiration

Ancestral gill books have been replaced by 'lung books' (also called ' book lungs') where perhaps as many as 750 leaflets full of blood are separated by air spaces (Figure 14.4). Many arachnids have also developed tracheal systems, resembling those of insects except that in arachnids they convey oxygen not to every cell but to the blood, where it is carried round the body by the respiratory pigment haemocyanin. Spiders may have various forms of tracheae as well as lung books; some have tracheae alone. The tracheae open by spiracles, which can be closed to restrict respiratory water loss. The most highly developed tubular tracheae occur in the fast-running solifugids, where air in the tubules is moved by a degree of muscular ventilation.

The arachnid heart lies dorsally in the opisthosoma. Primitively blood enters it by many ostia, but where tracheae take over from lung books the heart becomes simpler; it has only two pairs of ostia in small tracheate spiders. The heart pumps the blood forward by way of a series of dorsal arteries to the haemocoel of the prosoma, where increase in blood pressure protrudes the walking legs: the blood pressure may double when jumping spiders jump. The constricting waist of spiders allows the pressure in the prosoma to be independent of that in the opisthosoma, and in solifugids the two circulations can be separated by closure of a diaphragm. That the blood serves both to transport oxygen and to extend the legs can limit the activity of arachnids: if forced to sustain activity they become short of oxygen.

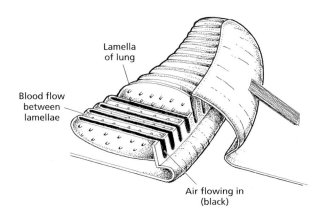

Lamella of lung

Blood flow between lamellae

Air flowing in (black)

Fig. 14.4 The book lung of a spider.

14.5.3 Excretory organs

The coxal glands serve for osmotic regulation. Study of a tick reveals ultrafiltration into a sac, sucked by external muscles. Nitrogen excretion occurs not in the coxal glands but in 'Malpighian tubules', analogous to those of insects (see Chapter 15), or in the walls of the midgut. The nitrogenous end point is guanine, a purine even more insoluble than uric acid.

14.5.4 Sense organs

In arachnids, sense organs are of three kinds: hairs on the cuticle, modified to detect mechanical or chemical stimuli; eyes, usually simple rather than compound and often spread over the body; and slit sensory organs, pits in the cuticle each covered by a thin membrane bulging inward towards a sensory hair, able to detect slight changes in the tension of the cuticle. These last are used largely as proprioceptors, i.e. to inform the animal about the degree of contraction of its own muscles.

14.5.5 Reproduction

Fertilisation is internal, by direct transfer in harvestmen but more usually by way of a package of spermatozoa done up in silk. These packages are deposited on the ground by scorpions, pseudoscorpions and some mites, but spiders (uniquely) handle the package with their pedipalps and, after ritualised courtship, insert it into the female. Courtship allows the male to avoid being mistaken for prey: in some species he risks being eaten by the female after mating but, contrary to popular belief, this is rare. Development is direct, in large eggs with the yolk central and uncleaved, as in insects.

14.5.6 Silk

Silk is a protein, fibroin, containing an unusually high proportion of the amino acids glycine and alanine. It is extremely strong: ten times stronger than collagen, as strong as nylon but more elastic. Silk is unique to arthropods, and has evolved a number of times, including in insect larvae and in some myriapods, but it is particularly important in spiders. The silk glands include a tubule in which the raw material of silk is made more acid and water is withdrawn to improve gel formation. The glands are at the posterior end of the opisthosoma, opening by mobile spinnerets; these and the mobility allowed by the narrow waist enable the spider to place the silk, which is not squirted into the air but attached to a surface. The silk is then hardened because the weight of the spider pulls out the thread, which alters its molecular configuration. This process can be checked by reduction of pressure in the opisthosoma.

Spiders use silk for sperm transfer, egg cases, nest linings, drag lines and for dispersal: 'ballooning' spiders can drift over long distances. Probably it evolved initially as a solution to reproductive requirements, but most characteristically it is used for food capture. The 'orb' web of the garden spider, characteristic of the large family

Araneidae, is made of two kinds of silk, one constituting the framework and the radial threads, between which is wound a spiral of the other kind, 'viscid' silk bearing glue droplets. Solid connections form wherever the threads cross. The viscid silk is extensible enough to catch quite large insects (not too large, they might be fierce). The spider sits at the centre or runs over the web (it does not stick to its own web partly because the legs secrete an oily substance and mainly because the principal 'spokes', along which the spider runs, are not sticky). Web spiders use their eyes very little. Their mechanoreceptors are extremely sensitive to disturbance of the web, and their slit sense organs are grouped as 'lyriform organs' that enable spiders to judge both the angle of change in tension in the cuticle and the distance they have covered over the web.

Different spiders make differently shaped webs, not necessarily two-dimensional: funnels, triangles, domed, sheets, placed horizontally or vertically. An elaborate web can be constructed within an hour, often only after the old web has been eaten. Spiders given caffeine or amphetamines make odd webs, deformed in ways constant for the particular drug. Young spiderlings make perfect webs without having to learn, yet there is some plasticity: remarkably, despite the role of gravity in normal web construction, spiders in space capsules can make perfect webs. Cursorial spiders (wolf and crab spiders, etc.) and jumping spiders have highly developed eyes and hunt by sight, often as ambush predators. They produce a drag line of silk, and may use it as a snare.

In summary, the production and use of silk is the key to the great success of spiders as predatory terrestrial animals.

MYRIAPODA

14.6 What are myriapods and how do they move?

Behind the head, a myriapod consists of a series of similar segments, each bearing one or two pairs of walking legs, with no division into regions such as thorax and abdomen, prosoma and opisthosoma (Figure 14.5). The head is distinctive: behind the first segment (without any appendages) and the second segment (having a single pair of antennae), the mouthparts are modified to form an upper lip, mandibles and a lower lip (labium) within which a buccal cavity is enclosed. Except in one group of centipedes, there are no compound eyes. All known myriapods are terrestrial. They comprise:

14.6.1 Chilopoda
Centipedes or Chilopoda are predatory carnivores, with the first pair of walking legs forming poison claws. About 3000 species are known, and they are up to 270 mm in length. Their soft bodies are flattened

dorsoventrally with the legs projecting laterally; they can insert their flat bodies into crevices and spaces in litter. 'A hundred legs' is not an accurate description: there may be many fewer, or as many as 177 pairs, all alike. Some run very fast, with a metachronal rhythm similar to that of polychaetes (Figure 14.5a and Chapter 9), with the wave passing from back to front. Tripping over the legs is a hazard: long and short legs may alternate, or more posterior legs may be longer and overlap (*Scutigera*, Figure 14.5b). The fast-moving scutigerids are the only myriapods to have compound eyes and respiratory pigments.

14.6.2 Diplopoda

Millipedes or Diplopoda feed on plants or decaying material. Their bodies are usually round in cross-section with a hard cuticle (Figure 14.5c). The body is held straight and the legs are

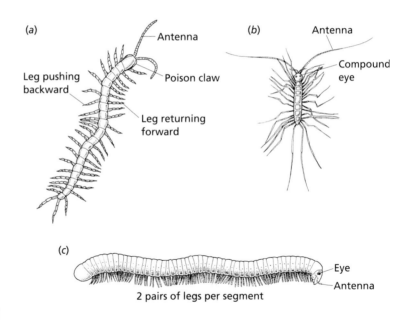

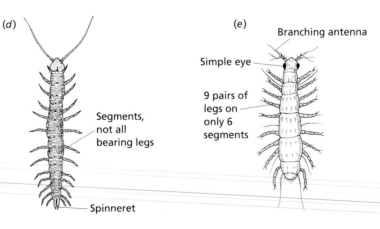

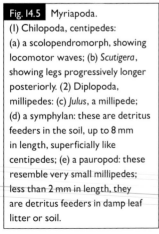

Fig. 14.5 Myriapoda.
(1) Chilopoda, centipedes:
(a) a scolopendromorph, showing locomotor waves; (b) *Scutigera*, showing legs progressively longer posteriorly. (2) Diplopoda, millipedes: (c) *Julus*, a millipede; (d) a symphylan: these are detritus feeders in the soil, up to 8 mm in length, superficially like centipedes; (e) a pauropod: these resemble very small millipedes; less than 2 mm in length, they are detritus feeders in damp leaf litter or soil.

ventral; unlike the fast-running centipedes millipedes characteristically walk slowly, with a powerful thrust. They can push themselves into litter and decaying logs and most are able to bend dorsoventrally and curl their bodies into spirals. About 10 000 species are known, up to 280 mm in length, living in dark humid places. They do not have a thousand legs, but each apparent segment is double and bears two pairs of legs.

14.6.3 Symphyla and Pauropoda

The main characteristics of Symphyla and Pauropoda are indicated in Figure 14.5d,e. The interrelationships of these groups are controversial, except that the Symphyla are close to centipedes and Pauropods to millipedes. Myriapods as a whole may not be monophyletic. They are not now thought to be particularly closely related to insects (see Chapter 20).

14.7 How well are myriapods adapted to life on land?

How myriapods invaded the land is not known; possibly by the interstitial route between sand grains. The earliest fossils found, as late as the Silurian, are in terrestrial deposits and there is nothing to indicate a freshwater origin. The cuticle is not very impermeable compared with that of arachnids, often containing little wax. Excretion is by Malpighian tubules, less highly developed than those of insects. Respiration is by tracheae; unlike insect tracheal systems they do not convey oxygen to each cell but to the blood, as in arachnids. The pattern of tracheal systems is related to the level of activity: millipedes have spiracles on all segments behind the third and short tufts of tracheae, while centipedes, more active, have fewer spiracles but a pair of lateral tracheal trunks giving off branching tubes, as in insects. In many species the spiracles cannot be closed and are therefore an important limit to water conservation.

In summary, myriapods are only incompletely terrestrial, with most species confined to damp habitats. For arthropods even better adapted to land life than the spiders, one must look to the insects.

Chapter 15

Insecta

Most animals are insects and one group, the beetles, is the largest known order of animals. There are vast numbers of individuals, over a million described insect species and many times that number awaiting description. They are by all criteria outstandingly successful, primarily on land, where the close relationship between insects and flowering plants first evolved, but many species occur in fresh water, often as immature stages. Unlike crustaceans, there are few species in the sea. The following account first explains that insects owe their success as terrestrial arthropods to a number of special characteristics, such as their water conservation mechanisms (already introduced in the previous three chapters) and the power of flight, which is so important that it demands a relatively full discussion. The life cycle is introduced to show that it provides flexibility, especially in those insects where there is a pupa, a transitional form that may provide a resting stage additional to the egg. An indication of the range of insect orders is given, the evolution of social insects is briefly discussed and finally the contribution of the fruit fly *Drosophila* to our knowledge of genetics is explained.

15.1   What is an insect?

Insects typically have three pairs of legs, and accordingly are called 'hexapods'. The adult body is divided into head, thorax and abdomen. The head bears a pair of compound eyes. The paired segmental appendages are one pair of antennae, typically mandibles, maxillae and a lower lip or 'labium' (in place of second maxillae) on the most posterior head segment. The form of the mouthparts varies greatly according to diet. Each of the three thoracic segments bears a pair of legs, and typically there is a pair of wings on each of the two posterior segments. The abdomen has 9 to 11 segments and no appendages (except genitalia or their derivatives). It bears the gonopores and the gut opens at the posterior tip. Excretion, including water regulation, is carried out by Malpighian tubules at the junction between the mid and hind gut. Respiratory gases are transported between the outside air and every cell of the body by tracheae, branching tubes lined with cuticle that open at the body surface by holes called spiracles. The small size of insects (the largest present-day species approach the size of a mouse and most are very much smaller) is dictated partly by the requirements of the tracheal system but probably even more by the difficulty of supporting the body in air when the exoskeleton is shed at moulting. Insects usually dominate ecological niches requiring small size.

15.2 Why are insects such successful land animals?

Small size is a crucial attribute of insects. Competition with vertebrates can be avoided by animals two orders of magnitude smaller: ecological niches also are related to size, and insects have been able to dominate those that require small-sized occupants. A further advantage of small size is rapid reproductive turnover, which not only accelerates population increase but also allows faster evolutionary change. The outstanding physiological advantage of being small is that the tracheal system can supply oxygen directly to every cell of the body (Figure 15.1a). Air enters the tracheae through thoracic and abdominal spiracles and diffuses (aided in the larger insects by ventilation movements) along the much branched and anastomosing tracheal tubes to the thin-walled 'tracheoles' which penetrate between or even into cells (Figure 15.1b). Unlike arachnids, insects rely totally on the tracheal system for transport of oxygen: the blood plays no part in respiration and contains no respiratory pigment. (As always with insects, generalisation is rash: there are some midge larvae, 'blood worms', containing haemoglobin.) The tracheae are lined by cuticle, with spiral thickening that keeps the tubes open. At moulting all tubes except the tracheoles have to be shed completely and replaced.

Small size, however, has the overwhelming disadvantage that the surface area of the animal is very large in proportion to the volume, so that on land there is great risk of desiccation. The success of insects depends upon water conservation, which is achieved to an unprecedented extent.

15.2.1 The cuticle
The cuticle (see fuller account in Chapter 12) is almost completely waterproof, due to the organisation of the wax layer. An ordered molecular structure is suggested by the effect of temperature change on the permeability of the epicuticle, but the details are still controversial.

15.2.2 The tracheal system
No other animals can take oxygen from the air with so little accompanying water loss. Only in the tracheoles are gases exchanged

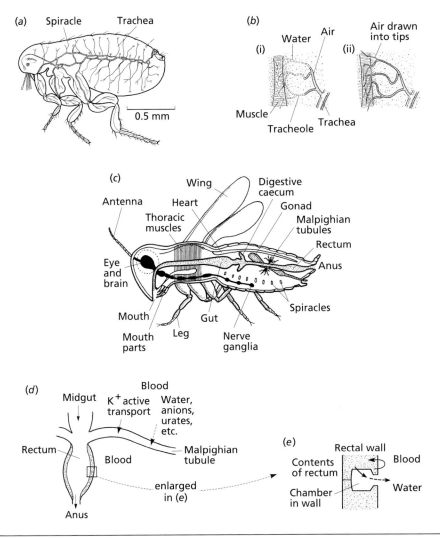

Fig. 15.1 Insect tracheal systems and Malpighian tubules: (a) one half of the tracheal system of the flea, *Xenopsilla*; (b) tracheoles supplying a muscle fibre, (i) at rest and (ii) when active; (c) diagrammatic section through a generalised insect, showing the Malpighian tubules; (d) diagram of a Malpighian tubule in relation to the blood and the rectum; (e) diagram showing ion and water movements within the rectal wall. Solid arrows denote active ion transport, dotted arrows denote passive movements of ions or water.

over a damp surface where water could evaporate, and the tracheoles are far removed from the body surface. The distant spiracles are the only sites of water loss from the body, and they are kept closed by valves for much of the time.

15.2.3 The Malpighian tubules and the rectum

These organs are uniquely efficient at regulating water loss during excretion of metabolic waste in the very insoluble form of uric acid. Malpighian tubules are long thin tubes arising at the junction of the mid and hind gut and extending into the haemocoel, where they are bathed in blood (Figure 15.1c).

Most insects have no appreciable blood pressure. Therefore waste cannot enter the excretory tubule by ultrafiltration, as occurs in most other animals, including most groups of arthropods. Instead, ions are actively transported from the blood into the upper part of the tubule; often only one kind of ion is moved actively and ions of the opposite charge follow passively (Figure 15.1d). Potassium is frequently the ion moved actively; chloride follows, with many other anions including urates (the salts of uric acid), amino acids, sugars and other metabolically useful substances. The resulting osmotic potential draws water also into the tubule and the solution passes down to its base where it enters the rectum. Here the pH is lower, which causes uric acid to crystallise out. The rectal walls are very muscular and are protected by cuticle from toxins and from abrasion. Some part of the wall is thickened to form compartments into which ions are actively transported (Figure 15.1e). Water from the contents of the rectum is then also drawn into those compartments and passes on through open channels into the blood. Metabolically useful substances are similarly reabsorbed. The rectum then squeezes out undigested food and uric acid crystals through the anus, with hardly any accompanying water. Some insects even absorb water from the unsaturated atmosphere through the rectum.

Malpighian tubules are under hormonal control, and their rate of ion secretion can be altered if water needs to be removed rather than conserved. For example, when the bloodsucking bug *Rhodnius* has had a large meal, the resulting stretching of the abdomen stimulates nerve endings in its wall and these nerves cause cells in the middle thoracic ganglion to release a diuretic hormone into the blood. This can increase the rate of excretion a thousand-fold, and water is lost at the anus until equilibrium is restored.

In summary, the tracheal system, the uniquely thorough removal of water from faeces and excretory material, and above all the waterproofed cuticle, enable insects to be outstandingly successful terrestrial animals. The almost complete absence of insects from the sea should not surprise us: their elaborate adaptations to life on land would there be irrelevant or disadvantageous (for example, a small air-filled arthropod could not dive to escape predation), and they could not withstand the competition from copepods in the plankton.

15.3   How are insects able to fly?

Insects are the smallest flying animals. The essential requirements for flight are the same as those for birds: they need wings providing lift, an upward force opposing gravity, and thrust, a forward force opposing the backward drag which inevitably accompanies lift.

While the essential requirements are the same, the ways in which flying animals obtain lift and thrust depend greatly upon their size.

15.3.1 Insect wings

The wings are paired extensions of the cuticle of the dorsal part of the second and third thoracic segment. These flaps of cuticle contain haemocoel and tracheae (Figure 15.2a). To expand the wings initially (at the final moult) blood is pumped into the haemocoel: it then dries out and nearly all the epidermal cells die, leaving the wings very light and strong.

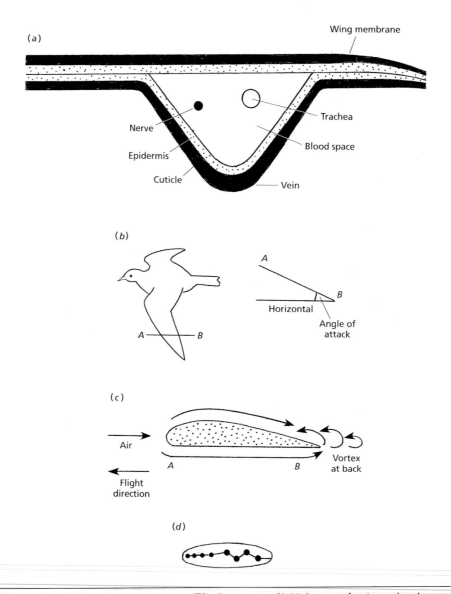

Fig. 15.2 Insect wings and flight: (a) transverse section (TS) of insect wing; (b), (c) diagrams of a wing to show how an aerofoil may obtain lift; (d) TS insect wing with boundary layer; (e) air movements in (i) clap, (ii) fling, (iii) end of fling; (f) the leading-edge vortex.

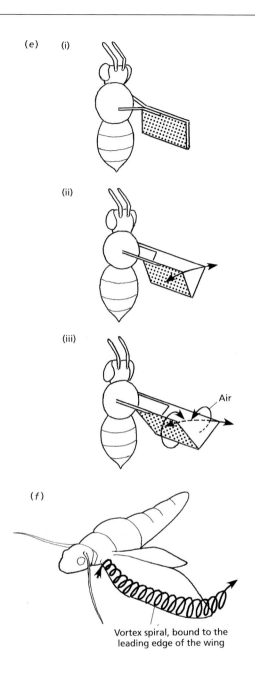

Fig. 15.2 (contd.)

Vortex spiral, bound to the
leading edge of the wing

15.3.2 Lift

Lift is provided in any wing, and accompanied by thrust, when the
air moves faster over the upper surface than over the lower,
provided also that the 'angle of attack' is positive (Figure 15.2b,c).
A bird's wing in cross-section has the shape known as an 'aerofoil
section', the curved upper surface providing a longer path for the
air flowing over it. The airstream over the lower surface therefore
reaches the back of the wing first, and air is sucked upward to

meet the airstream over the upper surface. A vortex is set up at the back (the 'trailing edge'), pulling towards it the air over the upper surface so that it moves faster than the air over the lower surface, providing lift. A large bird can glide: a large wing provides a great difference between the rates of air flow over the upper and lower wing surfaces. In large flying animals, flapping is not essential to obtain lift.

The small size of insects demands a rather different mechanism, as detailed below.

Flapping: an insect, like a small bird, must flap to obtain sufficient lift, by increasing the rate of air flow over wings moving much faster than the main body (lift is proportional to the square of the velocity). The small size of insects is favourable for flapping, because the power-to-weight ratio is high.

The boundary layer of inert air sticking to the wing surfaces constitutes the greatest problem for a very small flier. Owing to the viscosity of air any wing has a boundary layer, of the same width for any size of wing: a boundary layer of 1 mm is negligible in birds, but relatively enormous in insects (Figure 15.2d). In consequence of this large 'viscous drag', the lift-to-drag ratio (on which the speed and efficiency of flight depends) is at best only about a tenth of that in birds. Any increase in surface area increases viscous drag, and accordingly insect wings are flat plates, not aerofoil sections. However, an aerofoil section is produced when a flat plate with an adhering boundary layer is moved through the air, and this shape enables lift to be produced.

Obtaining sufficient lift remains difficult for a very small flier, as there is very little difference between air flow rates over the two wing surfaces, and to obtain sufficient lift small insects must flap their wings very fast indeed. Many insects turn their wings to get lift on the upstroke as well as the downstroke. Any increase in the rate of air movement over the upper surface compared to the lower surface will increase lift: for example small insects may bend the wing along its length and suddenly release it, or clap the wings together dorsally and then fling the leading edges apart (Figure 15.2e). Hovering, when there is no air movement provided by the whole body moving forward, is the most exacting form of flight. Lift is acquired only after several strokes and even then it is only obtained near the bottom of the downstroke and lost again as the wing turns for the upstroke.

Recent analysis emphasises that the forces acting on a wing that stops and starts repeatedly as it beats up and down are different from those on a wing moving at constant speed. Air over the leading edge forms a vortex bound to the upper surface of the wing and spiralling out towards the wing tip (Figure 5.2f). This is the main aerodynamic mechanism enabling small insects to fly.

15.3.3   Methods of flapping

Insect wings are raised by contraction of indirect dorsoventral muscles that pull down the roof of the thorax (Figure 15.3a). For the wing downstroke, muscles are of three kinds:

Direct muscles attached to the wing bases, pulling them down. These occur in dragonflies but in few other insects.

Indirect muscles. In most insects, changes in shape of the thorax also move the wings on the downstroke, by contraction of the anteroposterior muscles (Figure 15.3b).

Asynchronous ('myogenic') muscles. These are indirect muscles found only in Diptera, Hymenoptera and some small beetles and bugs. This form of striated muscle is able to contract much faster than nerves can conduct, being stimulated by mechanical stretch, usually caused by an antagonist muscle. The muscles depend on nerve impulses for background stimulation, and rely on energy stored in resilin in the wing hinge. These are the fastest, most elastic and most isometric muscles known, enabling up to over 1000 wing beats per second.

15.3.4   Control of flight

The wing-flapping rhythm is set by groups of interneurons constituting the 'central pattern generator', situated in the thoracic ganglia. The brain receives sensory input from eyes, antennae and mechanoreceptors placed at the base of wings, the whole head (dragonflies) or in Diptera the halteres, modified hind wings which beat up and down with the fore wings, any asymmetry of strain being detected by the receptors at the haltere base. The brain receives and coordinates the sensory input and may change the flight rhythm. Correction cannot be instantaneous in any insect, since the wing moves too fast, and the only changes that the insect can make are to alter the angle of attack or the stroke plane of the wing, or to stop flying.

15.3.5   The metabolic cost

The energetic cost of flight is enormous. Oxygen is never limiting for an insect: the tracheal system can supply enough oxygen sufficiently fast to fuel the smallest flapping insect. Oxygen diffuses faster in air than in water by a factor of 10 000, and with a primary tracheal trunk up to 0.74 mm long (as in locusts) no energy need be spent on ventilation movements. Tracheoles may even indent muscle fibres (Figure 15.3c) which can be as large as 500 μm in diameter. By contrast, activity in vertebrates is limited by the rate at which oxygen can be supplied from blood capillaries (the diameter of the muscle fibres does not exceed 20 μm, exactly as in dragonflies, where there is no tracheolar indentation). Vertebrates often need to draw on anaerobic respiration, where food is oxidised only as far as lactic acid formation and an oxygen debt is built up. Insects never need to be anaerobic (and lactic acid would flood the small body). Food supply is the limiting factor for an active insect. The blood that supplies food

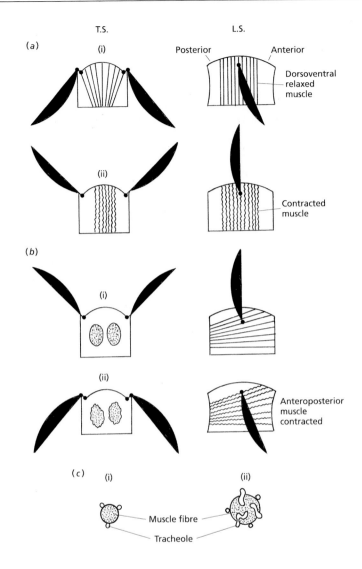

Fig. 15.3 Insect flight muscles in transverse section (TS) (left-hand column) and longitudinal section (LS) (right-hand column): (a) in upstroke, for all insects, (i) before and (ii) after muscle contraction; (b) in downstroke, for most insects, (i) before and (ii) after muscle contraction; (c) TS muscle fibres with tracheoles, (i) in a dragonfly and (ii) in a fly.

is not circulated fast, and muscle contraction squeezes it away from the sites where it is needed. However, insect blood can contain large and variable quantities of food. It is 1–3% carbohydrate, with similar quantities of fat and about 2% amino acids. The sugar is mostly trehalose, a disaccharide that can be tightly packed, rather than the monosaccharide glucose as in vertebrates. Mammalian blood by contrast contains small and precisely regulated quantities of foods (0.1% sugar and 0.1% fat, with very little amino acid except in the vessel between the intestine and the liver). This is necessary to maintain the stability of the respiratory pigment haemoglobin, not present in insects.

Only adult insects can fly (but see below: mayflies). Some adults are wingless or seldom fly, and all other stages in the life cycle rely on other means of locomotion such as walking, crawling, swimming or burrowing.

15.4   What is distinctive about insect life cycles?

The egg is often very resistant and may be an important resting stage. Its early development is described in Chapter 19. Very often there is division of labour within the life cycle, made possible as in crustaceans by metamorphosis: many insects have a feeding larva different in form and food supply from the reproductive adult. Unlike crustaceans, however, there is often a winged adult, and this rather than the larva is the agent of dispersal.

15.4.1   Exopterygotes

These are the insects that have a gradual metamorphosis from the larva (sometimes called a 'nymph') into the adult ('imago'). Wing buds grow externally on the larval thorax (hence 'Exopterygota'). Many exopterygote adults and larvae, such as those of cockroaches, look very similar, live in the same environment and presumably compete for resources. Aquatic larvae (as in stoneflies and dragonflies) differ greatly from the adults in feeding mechanism and appearance. In mayflies the aquatic larva lives for up to three years and the very short-lived aerial adult has no gut and cannot feed at all.

15.4.2   Endopterygotes

These have 'complete metamorphosis', with total reorganisation of the body within the pupa. From the outset the larva contains the rudiments of adult organs (the 'imaginal discs'). These remain inactive during larval life but at the final moult, when, in the absence of juvenile hormone, the pupa is formed, the adult rudiments develop and the larval tissues are resorbed. Endopterygote larvae typically look very different from the adults, have different feeding methods and exploit different niches.

15.4.3   Diapause

Stages of inaction in the life cycle are often necessary to match an animal to its seasonal environment. The pupa of endopterygote insects may appear to be a resting stage while there is great activity inside, but then it may become entirely quiescent and provide a resting stage in the life cycle, additional or alternative to the egg. 'Diapause' is a period of arrested development, characterised by low oxygen consumption. The cue to enter diapause is most frequently a change in day length ('photoperiod'), rather than in temperature or humidity, which may be very variable. In the cabbage white butterfly (*Pieris rapae*), for example, the shorter day length at the end of summer switches off brain hormone in the final larval stage. The pupa can be formed, but cannot moult, and therefore goes into diapause. After an extended period of chilling (winter) the brain hormone switches on again, the pupa moults and the butterfly emerges.

15.5   What are the main orders of insects?

The enormous numbers of insect species are assigned to many orders, usually differing in life histories and in characteristic feeding methods. Like the different phyla of the Animal Kingdom, orders are distinguished by recognisable basic forms (of adult or larva) showing adaptation to particular niches. There are two major divisions of insects:

(I) **Apterygota**: some insects have lost their wings, but Apterygota are primitively wingless hexapod arthropods, an assemblage of forms not closely related (Figure 15.4a). Possibly these evolved separately from each other and from Insecta; in particular springtails (Collembola) may be an entirely separate group of terrestrial arthropods.

(II) **Pterygota**: insects primitively winged. There are two groups:

1. **Exopterygota** (Figures 15.4 and 15.5), with gradual 'incomplete' metamorphosis and the wings developing externally as larval wing buds.

(A) **Palaeoptera** are those exopterygotes in which the wings are not folded over the abdomen when at rest but are held out sideways or up above the insect's back. There are two present-day orders, both with aquatic larvae:

Odonata (Figure 15.4b): dragonflies, active predators with biting mouthparts, small antennae, very large eyes and a long abdomen extending straight out posteriorly. The wings at rest are held vertically in the delicate 'damselflies' and horizontally in the

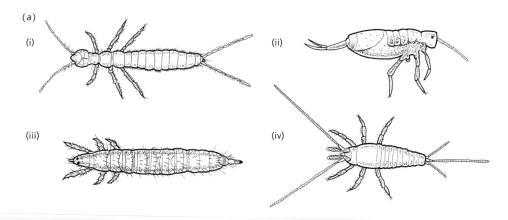

Fig. 15.4   The diversity of insects: apterygotes and palaeopteran exopterygotes (Odonata and Ephemeroptera). (a) Apterygotes: (i) dipleura; (ii) springtail (length c. 1 mm); (iii) protura (length 7 mm); (iv) bristletail (length 20 mm). (b) Anisopteran dragonfly, (i) adult and (ii) aquatic larva; (c) two damselflies in tandem; (d) mayfly, (i) adult and (ii) aquatic larva.

(b)

(i)

(ii)

External
wing buds

Fig. 15.4 (contd.)

(c)

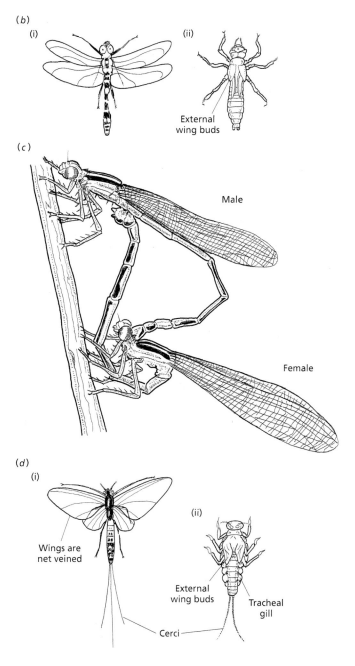

Male

Female

(d)

(i)

Wings are
net veined

(ii)

External
wing buds

Tracheal
gill

Cerci

larger true 'dragonflies' (the name may be reserved for this subdivision or given to the whole order). They are ancient insects whose uniform structure has changed remarkably little since the Carboniferous. Fossil insects from this era with a wing span of about 0.6 metres are the largest insects known. Their mating system is unique (Figure 15.4c). The male usually patrols a territory beside water, captures a female and flies 'in tandem' with her. He fills his accessory genitalia under the front end of the abdomen with sperm, cleans out from the female any sperm from a previous mating, and then inserts his own sperm. Eggs are laid in water or on submerged plants. The aquatic larva seizes prey with a 'mask'; it may live in water for several years.

Ephemeroptera: mayflies. Figure 15.4d shows the adult and larva. There is also a pre-imago stage which (unlike all other immature insects) has wings.

(B) **Neoptera** are those insects in which the wings are normally folded over the dorsal surface of the abdomen when at rest. They include:

Dictyoptera (Figure 15.5a): cockroaches and mantids.
Orthoptera (Figure 15.5b): locusts, grasshoppers and crickets.

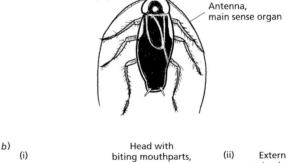

(a)

Antenna, main sense organ

(b) (i)

Head with biting mouthparts, herbivore

(ii)

External wing buds

Long hind leg (jumping, sound production)

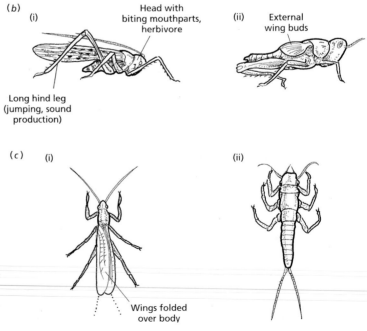

(c) (i)

(ii)

Wings folded over body

Fig. 15.5 The diversity of insects (continued): neopteran exopterygotes. (a) Dictyoptera, cockroach adult (larva, very similar, not shown); (b) Orthoptera, locust, (i) adult and (ii) larva, showing externally developing wing buds; (c) Plecoptera, stonefly, (i) adult, found on stones near water, and (ii) aquatic larva; (d) Dermaptera, earwig, (i) adult and (ii) larva; (e) Isoptera, termite, worker (soldier); (f) Hemiptera, bugs, adults of the two suborders, feeding by piercing or sucking plants or animals.

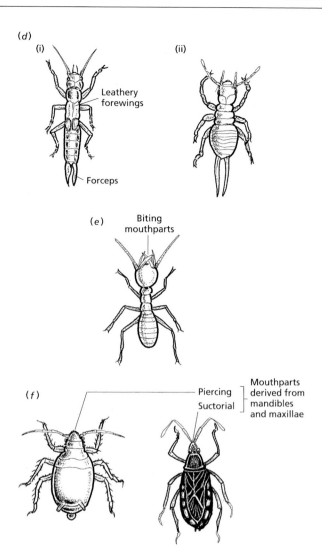

Fig. 15.5 (contd.)

Phasmida: stick insects and leaf insects, with highly adapted forms, as their names indicate. By day these nocturnal predators are effectively concealed among twigs and leaves.

Plecoptera (Figure 15.5c): stoneflies.

Dermaptera (Figure 15.5d): earwigs, which are semi-social and make good mothers – and do not inhabit the ears of any animals.

Isoptera (Figure 15.5e): termites, social insects sometimes called 'white ants', being colonial and polymorphic. They feed on wood and may be very destructive to buildings; they construct and inhabit the large often pointed mounds seen in the tropics.

Hemiptera (Figure 15.5f): bugs, a term often misapplied to other forms.

There are other smaller orders within Neoptera:

Anoplura: sucking lice.

Mallophaga: biting lice, ectoparasitic on birds and mammals.

Psocoptera: book lice, feeding on fungi, for example those growing on the bindings of damp books.

Embioptera: web-building insects living under bark in tropical countries.

Thysanoptera: thrips or thunderflies, tiny insects fringed with hairs. The mouthparts are asymmetrical.

Grylloblattoidea, which have been thought to be remnants of the primitive stock from which Dictyoptera and Orthoptera evolved.

2. **Endopterygota** (Figure 15.6), with complete metamorphosis and a pupal stage, within which adult structures including wings develop from imaginal discs. There are four major large orders and a number of smaller ones:

Lepidoptera (Figure 15.6a): butterflies and moths. A very large and well-known order of insects, united by having scaly wings. Lepidoptera are among the most recent insects: they arose in the Tertiary and evolved in close relationship with flowering plants. The larvae (caterpillars) feed on leaves, stems or roots and the adults feed by sucking nectar from flowers. Large diurnal butterflies owe their conspicuous and often beautiful coloration to the scattering of light by the scales on their wings. Most families of Lepidoptera are small nocturnal moths. Some primitive 'Microlepidoptera' feed using biting mouthparts, but in all others the mandibles disappear and the maxillae are enormously extended to form a sucking 'proboscis', coiled under the thorax when not in use.

Hymenoptera (Figure 15.6b) have small hooks joining the posterior edge of the fore to the hind wing on each side ('married' wings). This diverse order includes:

(a) **Symphyta**, sawflies, with larvae looking like caterpillars and adults with no obvious 'waist'.

(b) **Apocrita**, with grub-like larvae and adults with a constriction between the thorax and abdomen. These include:

(i) **Parasitica**: ichneumonid flies, gall wasps, which lay eggs in other insects or in plants; the larvae live as parasites inside these hosts (and may be important regulators of the population sizes of some of our pest species).

(ii) **Aculeata**: bees, wasps and ants, with adults larger than in Parasitica and with the female ovipositor modified as a sting. All provide food for their helpless larvae, in the form of pollen and honey (bees), insect prey (wasps) or fungi cultivated on plant material (some ants). There are many species of solitary bees and wasps but most of them, and all ants, are social insects (see below).

Coleoptera (Figure 15.6c) beetles are characterised by 'elytra', fore wings modified into hard resistant cases under which the membranous hind wings are folded. When not in flight a beetle appears to be completely covered in armour. All beetles have strong biting mouthparts. Beetles are first found in the Permian as a very compact and isolated group of insects; today they constitute a uniquely vast and diverse order, with about half a million species ranging from 0.5 to 210 mm in length. The larvae are typically legless grubs, but some are predators with jaws and legs.

Diptera (Figure 15.6d): true flies have only one pair of wings. The hind wings have become modified into halteres, which are sensory structures assisting flight (see above). Flies are relatively recent insects that arose in the Cretaceous, in association with flowering plants. There are very many diverse species including insects with piercing and sucking mouthparts (e.g. midges, mosquitoes, craneflies, horseflies) and the great majority (e.g. house flies, fruit flies) which have mouthparts specialised for sucking only. Many dipteran larvae are headless maggots living in soft substrata.

Smaller orders of endopterygotes include:

Trichoptera: caddis flies, a small group resembling moths but with hairs on their wings rather than scales. There is a prolonged aquatic larval stage, when the larva usually builds a characteristic case of assorted fragments of stones etc. from which its head projects for feeding.

'Neuroptera' (Figure 15.6e): alder flies, lacewings and ant-lions are not a natural group but a combination of endopterygote orders where both adults and larvae are predators with biting mouthparts. The wings look like nets of gauze and are held tent-like over the abdomen; flight is rather feeble.

Mecoptera: scorpion flies, first found as fossils in the Lower Permian, are thought to be survivors of the primitive stock from which evolved the 'Neuroptera', Lepidoptera, Trichoptera and Diptera.

Siphonaptera (Figure 15.6f): fleas, laterally compressed and wingless ectoparasites that pierce the host with stylets and suck blood.

Strepsiptera (Figure 15.6g), much modified parasites of some bugs and Hymenoptera.

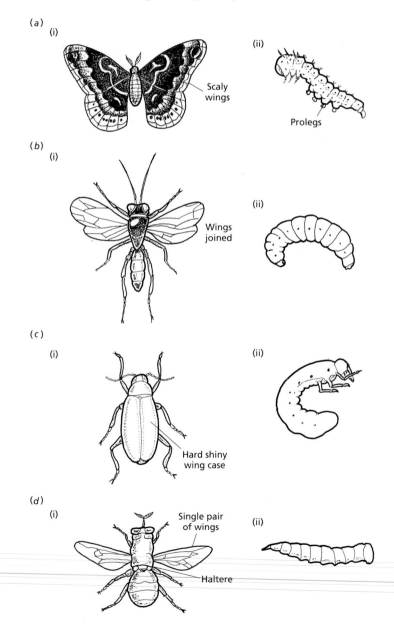

(a)
(i)
Scaly wings
(ii)
Prolegs

(b)
(i)
Wings joined
(ii)

(c)
(i)
Hard shiny wing case
(ii)

(d)
(i)
Single pair of wings
Haltere
(ii)

Fig. 15.6 The diversity of insects (continued): endopterygotes, generalised examples, in each case showing (i) adult and (ii) larva.
(a) Lepidoptera, butterfly;
(b) Hymenoptera, bee;
(c) Coleoptera, beetle; (d) Diptera, fly; (e) Neuroptera; (f) Siphonaptera; (g) Strepsiptera.

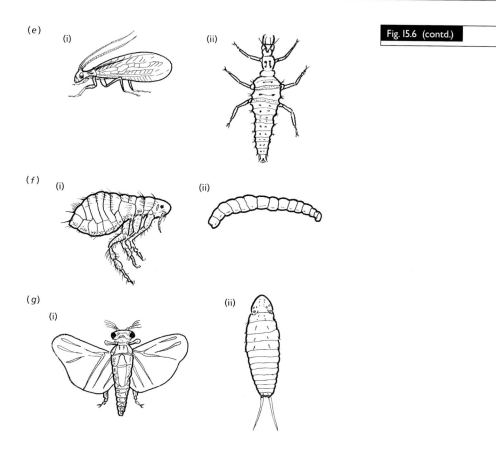

Fig. 15.6 (contd.)

15.6 How could social behaviour have evolved?

Ants, bees and wasps among the Hymenoptera (and termites, separately evolved in the exopterygote Isoptera) not only live in colonies but show remarkable division of labour among the members of those colonies. The biology of social insects is fascinating, but only one question will be considered here: how could this behaviour have evolved? This is not a problem about lack of intermediate stages: there are for example many solitary bees showing rudiments of social behaviour. The problem is to identify selective advantage for apparently altruistic behaviour. Typically there are very few reproductive individuals: hymenopterans have only a single queen per colony and the males (drones) die after the nuptial flight, during which she collects sperm that lasts for the rest of her life. The workers (and soldiers in ants) are all sterile females. In termites, each colony has a queen, who grows enormous, and a fertile king, who lives with her and fertilises her eggs as they are produced, and the workers are sterile males or females. How could it benefit the individual workers to cooperate in the rearing of their siblings and work to maintain the colony, instead of using their own reproductive potential? Natural selection acts on individuals, not on species as a whole.

The problem of altruism disappears when the focus is shifted to the genes. Natural selection acts upon individuals, but the effect of its action is that some groups of genes rather than others will be passed down the generations (see Chapter 1). If two individuals are closely related, the reproduction of either may be equally advantageous for transmission of a given gene. There is in this context nothing special about offspring. When siblings are as closely related to each other as to potential offspring, their genes are as likely to be transmitted when all siblings promote the reproduction of one of them. In Hymenoptera a special mechanism makes sibling workers even more closely related to each other than to potential offspring, thereby increasing the advantage of their social behaviour.

15.7 Why has study of the fruit fly *Drosophila* been so important?

Drosophila was T. H. Morgan's inspired choice for animal breeding experiments at Columbia University in the earliest years of the twentieth century. These fruit flies (*Drosophila melanogaster*) were already in 1901 being cultured for embryological work in the laboratory at Columbia. They are ubiquitous in the wild and easily obtained and cultured; they are small insects that produce large numbers of offspring in a very short time, and may complete their life cycle in two weeks.

15.7.1 Early work by Morgan and his group

Inbreeding and outbreeding experiments confirmed that Mendel's laws (see Chapter 1) applied to fruit flies as much as to garden peas. Work on *Drosophila* revealed the phenomena of linkage and crossing over, showed that genes were arranged longitudinally along chromosomes and related the position of genes to the frequency of the breaking of linkage. The job was relatively easy in *Drosophila* since there are only four pairs of chromosomes (later work used a different species, *D. pseudobscura*, where the chromosomes can be seen more clearly). *Drosophila* was a fortunate choice, since the salivary glands in the larva have giant 'polytene' chromosomes made from many strands of the genetic material in a bundle. These chromosomes were large enough to be accessible to experiment even in Morgan's time: under the light microscope transverse bands could be seen at different intervals, so that different parts could be recognised, as could 'puffs' when the genes were active. Genetic mapping was checked and extended by experiments in which parts of the chromosomes were cut out, turned round or rearranged.

T. H. Morgan, A. H. Sturtevant, H. J. Muller and C. B. Bridges in 1915 published *The Mechanism of Mendelian Heredity*. The merging of cytology and genetics into a single science was the first major contribution of the *Drosophila* work of the Morgan school.

15.7.2 The evolutionary role of mutations

Further work by Morgan's group, and by others such as Theodosius Dobzhansky and (with a gift of cultures from Columbia) S. S. Chetverikov in Russia in the 1920s, established the connection between Mendelism and Darwinism. At that time genetics and evolution were viewed very separately, with much controversy about their relationship. H. De Vries, working on evening primroses (*Oenothera biennis*), observed sudden large heritable changes that he termed 'mutations': these were generally deleterious and could not be seen as the raw material of evolution. Morgan redefined mutations as small changes in genes, initially recessive but retained in the genotype, and showed how such mutations could be the source of variation acted on by natural selection. For study of mutations *Drosophila* was a rewarding subject, having the complex insect structure with many characters under direct genetic control. A large 'library' of mutations was soon acquired, ranging from large-scale malformations to patterns of bristles on the cuticle. Mutations were shown to occur in wild populations also, to be imitated by heat shock at critical points in development (R. Goldschmidt) and to be caused by ionising radiation (Muller). A balanced polymorphism in the population was described by M. Teissier; the statistical approach to populations by R. A. Fisher was especially important. Genes were revealed as large and intricate entities which interacted with each other and with the environment, and genetics became part of the 'evolutionary synthesis' of the 1930s and 1940s.

15.7.3 Speciation

Speciation in *Drosophila* is marked, most spectacularly in Hawaii, where 450 species have been described. New Hawaiian islands are continually being formed and are biologically 'empty' at first. Small insects are weak fliers and colonists are few: these speciate readily as each island presents a great diversity of habitats, sexual selection is strong and lava flows may further divide populations. Many closely related species of *Drosophila* are therefore available for comparison; often there is remarkably little genetic difference between them.

15.7.4 Modern genetics

Accumulated knowledge of *Drosophila* species and mutants has provided an invaluable basis for application of modern knowledge and techniques concerning the nature and action of genes. Not surprisingly, *Drosophila melanogaster* was among the first invertebrates chosen for mapping of the entire genome.

Chapter 16

Animals with lophophores

Sessile animals require special structures for food collection: the stinging cells on the tentacles of cnidarians and the feeding methods of sponges and barnacles have already been discussed. The present chapter introduces four groups of animals with lophophores. As adults all are sessile, with poorly developed heads, U-shaped guts and a protective outer covering over what is most often a colony of zooids. A superficially similar phylum is also introduced for contrast, and the relationships of animals with lophophores are discussed.

16.1 What is a lophophore?

A lophophore is a circular or horseshoe-shaped fold of the body wall that encircles the mouth and bears numerous ciliated tentacles, the anus being outside the tentacular ring. The tentacles are hollow, each one containing a branch of a coelomic cavity separate from the main trunk coelom. Lateral cilia on the tentacles draw water bearing food particles into the tentacular ring, frontal cilia convey it down each tentacle and it goes out at the base, where food is trapped (Figure 16.1a).

16.2 Which animals have lophophores?

Three phyla, the Phoronida, Bryozoa and Brachiopoda, with lophophores exactly fitting the above description are frequently united as 'Lophophorata'.

16.2.1 Phoronida
Fourteen species are known, all marine, in two genera, *Phorona* (Figure 16.1a) and *Phoronopsis*, none more than 200 mm in length. The adult is solitary and worm-like, with unusual powers of regeneration. It is able to move only within the chitinous tube that

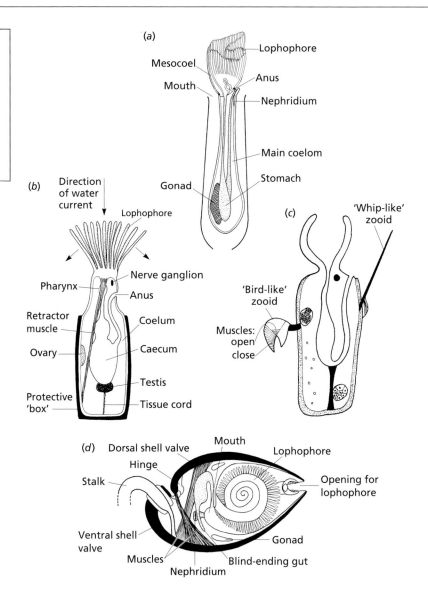

Fig. 16.1 Examples of three phyla with lophophores: (a) longitudinal section of *Phoronis*, showing the lophophore; (b) diagram of a generalised bryozoan zooid; (c) diagram of bryozoan with two greatly modified zooids superimposed; (d) a brachiopod with the lophophore coiled internally.

fastens it to the shallow sea bottom. The lophophore is either simple, with rather few tentacles, or elaborate and spirally coiled. There is a closed blood system with haemoglobin in corpuscles, the nervous system is subepidermal and largely diffuse. The excretory organs are protonephridial in the larva, metanephridial in the adult. The animals are mostly hermaphrodites, releasing gametes through the excretory pores, and the fertilised eggs develop into free-swimming 'actinotroch' larvae.

16.2.2 Bryozoa

The Bryozoa (Figure 16.1b) are very common intertidal animals. There are 20 000 species, of which 5000 are living today. They are the only phylum to be entirely colonial, with zooids about 0.5 mm long, and the largest class is marine, but there is one class that is entirely freshwater. Another is known mainly as fossils, very abundant since the late Cambrian. Bryozoa ('moss animals') have also been called 'Polyzoa' (because there are many individuals in a colony) or 'Ectoprocta' (because, as with all true lophophores, the anus is outside the tentacular ring).

Bryozoa are sessile colonies of miniaturised animals. Microscopic examination shows that each inhabits a separate little 'box' of cuticle within the colony, which at once distinguishes them from hydroid cnidarians, and their high speed of withdrawal into their boxes is a further distinction. The cuticle is made of chitin and protein, sometimes also calcium carbonate. The zooid body typically has a stationary trunk and an eversible lophophore, extended for feeding when muscular contraction increases the fluid pressure in the coelom and a special opening appears. There is a simple nerve ganglion giving off nerves, a U-shaped gut, and (in such small animals) no excretory or blood systems. The colonies grow by budding, or the zooids may contract down with regression of the gut and lophophore, leaving remnants as 'brown bodies' when most of the tissue regenerates. The zooids are hermaphrodites, the gametes being shed into the coelom and passing out through pores at the tentacular base. The fertilised eggs are usually brooded, and hatch as free-swimming larvae.

Bryozoa are diverse in form. Encrusting species grow as flat sheets covering seaweed, upright species grow vertically up from the substratum as tree-like tufts or 'bushes'. Division of labour between the members of a bryozoan colony commonly occurs. In addition to feeding zooids (described above) there may be 'bird-like' individuals with jaws serving for defence and 'whip-like' ones that clear away waste and deter settlers (Figure 16.1c).

The metabolism of Bryozoa shows great diversity: for example, in the Antarctic there are some species with the fastest metabolic rates known in polar animals and other species with the slowest. The great success of the phylum over evolutionary time is thought to be associated with such metabolic flexibility.

16.2.3　Brachiopoda

There are about 13 000 species, of which only 350 are living today. The shell is less than 100 mm long or wide. All are marine. Brachiopods are solitary sessile animals enclosed by the dorsal and ventral valves of a shell (Figure 16.1d). The two halves of the shell are unequal and the ventral half bears a stalk that anchors the animal to the substratum. Resemblance to the bivalve molluscs, where the two halves of the shell are lateral and equal, is only superficial. The main body of the brachiopod consists of a coelom surrounding the gut, excretory metanephridia and either testes or ovaries, as the sexes are separate. There is an open blood system and a ganglionic nerve ring round the oesophagus. Contraction of well-developed muscles moves the shell valves. The body lies posteriorly in the shell and the body wall extends forward, folded as a pair of mantles lining the shell. The horseshoe-shaped lophophore develops as a pair of arms that may become highly elaborate. It is coiled within the mantle cavity, from which it can be extruded.

Brachiopods are extremely common in the fossil record, from the early Cambrian onward. The stem groups reach their peak of diversity in the Ordovician and the greatest number of genera is found in the Devonian. They are later largely displaced by molluscs. Today they are still very widely distributed in the seas, although they are seldom common at any one place or depth.

There are two very different brachiopod groups, which may not be closely related. The Articulata (e.g. *Terebratula*) have calcareous shells with valves interlocking by teeth on one valve fitting into sockets on the other, and the dorsal valve usually has calcified outgrowths supporting the lophophore. The anus is lost in most living forms. The Inarticulata (e.g. *Lingula*) have organophosphatic shells with no articulation and no outgrowth. The valves are held together by complex musculature and there are other internal differences from Articulata. The earliest fossil brachiopods are inarticulate. *Lingula* is an outstanding example of a species which has persisted unchanged from the Ordovician to the present day.

16.2.4 Pterobranchiate Hemichordata

In addition to the above-mentioned three phyla, pterobranchs also have lophophores exactly fitting the above description, except that the tentacles do not completely encircle the mouth (Figure 16.2a) and the tentacular coelom is continuous underneath the tips of the tentacular horseshoe. Yet pterobranchs are members of the phylum Hemichordata.

16.3 Are animals with lophophores protostomes or deuterostomes?

This question is not easily answered, since phoronids, bryozoans and brachiopods present a mixture of the characters introduced in Box 5.2.

16.3.1 Phoronida

Like protostomes, the blastopore becomes the mouth, yet, like deuterostomes, cleavage is radial and early development is regulative. Further protostome-like characters are the protonephridial excretory tubules in the larva, the nervous system inside the

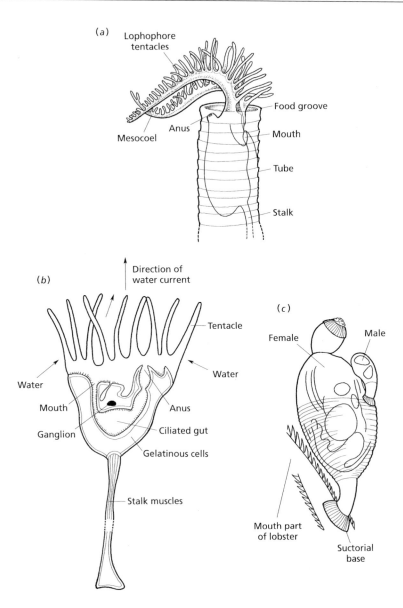

(a) Lophophore tentacles
Food groove
Anus
Mesocoel
Mouth
Tube
Stalk

Fig. 16.2 (a) *Rhabdopleura*, a pterobranch hemichordate; (b) a generalised entoproct; (c) *Symbion pandora*, a cycliophoran.

(b)
Direction of water current
Tentacle
Water
Water
Mouth
Anus
Ganglion
Ciliated gut
Gelatinous cells
Stalk muscles

(c)
Female
Male
Mouth part of lobster
Suctorial base

epidermis and the tube made of chitin. The mesoderm arises from the gut wall as in deuterostomes but the coelom is formed by splitting as in protostomes.

These contradictory characters have in the past been reconciled by placing phoronids as a basic group from which protostomes and deuterostomes have diverged, but further knowledge (including molecular evidence) disproves this.

16.3.2 Bryozoa

At first sight bryozoans are good deuterostomes, with radial cleavage and the blastopore forming the anus, although development is unusual in that the lophophore and gut bud from within the larva. Yet they have a number of resemblances to protostomes, notably the presence of chitin. Such resemblances might be dismissed but for molecular evidence: 18S rDNA analysis places bryozoans among the protostomes, not too distant from phoronids, annelids and molluscs.

16.3.3 Brachiopoda

Their development also does not fit neat categories. Like deutero-stomes, they have radial cleavage and regulative development. The blastopore forms the anus in some but not all species. Mesoderm formation, however, is enterocoelic only in the Articulata. In the Inarticulata, the more primitive group, mesoderm proliferates from the gut wall and then splits, just as in phoronids: yet another example showing that methods of coelom formation are more variable in evolution than has been assumed. Brachiopods, like bryozoans, are increasingly seen as protostomes, on account of both fossil and molecular evidence (see Chapter 20).

16.3.4 Pterobranchiates

These organisms remain classical undoubted deuterostomes by every criterion available – developmental, morphological and molecular.

16.4 What are the relationships of Entoprocta?

Entoprocta (Figure 16.2b) do not have lophophores, but are included here to highlight the distinctive features of those animals that do. There are 150 species of entoprocts, of length 0.5–5 mm. Most are marine with a few freshwater species. They may be solitary (e.g. *Loxosoma*) or colonial (e.g. *Pedicellina*). They are superficially similar to 'Ectoprocta' (Bryozoa), being small sessile filter-feeding animals, mostly colonial and marine. However, they lack a coelom, having instead many gelatinous cells in what might otherwise be a cavity. The tentacles are solid, borne on a collar which is not retractable but can be folded down, and the anus is within the tentacular ring. A further contrast with lophophores is that the water currents flow in the opposite direction, as lateral tentacular cilia draw water in at the base of the tentacular ring, up each tentacle and out at the tip, while frontal cilia trap food particles and convey them down to the mouth.

Entoprocts have a single protonephridium, no respiratory or blood systems, and nerves arise from a single ganglion. The animals are hermaphrodites and the egg hatches as a trochophore-like larva.

Entoprocts are undoubtedly protostomes, with spiral cleavage, the mouth formed from the blastopore and mesoderm developed from the 4d cells (it does not split, there is no body cavity). 18S rDNA analysis supports the placement of entoprocts among the protostomes, not particularly close to the bryozoans nor to other animals with lophophores.

The only phylum apparently close to the entoprocts was discovered in 1995 by P. Funch and R. M. Kristiansen: the phylum **Cycliophora** (Figure 16.2c), which is known only from one species, *Symbion pandora*, 0.35 mm in length, living on the mouthparts of the marine Norway lobster (*Nephrops norvegica*). The life cycle is complex. The sessile feeding stage has a funnel-shaped crown of tentacles, and water passes from the base to the tips of the tentacles as in entoprocts. The animal is solitary, enclosed in a sculptured cuticle resembling that of some nematodes, with a stalk for temporary attachment. This stage is female, with a small male attached. After fertilisation, development is that of a typical protostome and the larva is like a trochophore, except that internally it buds off a clone of individuals, which settle to form the feeding stage.

16.5   Should there be a group called 'Lophophorata'?

Should the animals with lophophores be combined in one group called 'Lophophorata'? Phoronida, Bryozoa and Brachiopoda may be fairly closely related, and present evidence establishes them as protostomes. However, the pterobranchiate Hemichordata are classical deuterostomes by every criterion available, developmental, morphological and molecular. This is most readily explained as a remarkable example of convergent evolution of the feeding structure. To combine the three closely related phyla as 'lophophorates' is frequently done but incomplete, and to include the very distantly related hemichordate example would be misleading. Accordingly, this chapter is called 'Animals with lophophores'.

Box 16.1 | How animals feed

Animals cannot manufacture organic foods from their inorganic constituents, and therefore they are dependent on green plants or microorganisms for their food. As land animals we are used to the assumption that our food chains begin with photosynthesis by green plants, but in the sea this is relatively rare. Originally the sea contained single-celled organisms only, and these are still the basis of many marine food chains. Light does not penetrate far and photosynthesis can occur only very near the surface; detritus falling from the surface layers becomes the basis of many food chains in deeper waters and on the sea bottom. Chemosynthetic bacteria are important, in particular the sulphur-based bacteria in deep sea vents (see Box 18.1). Small animals, or those with a large food-collecting area, can exist on a purely microbial diet.

Small particle feeding

Suspension or 'filter'-feeding

Filter-feeding can only occur in water. It is very common in marine larvae, and most phyla have examples of adult animals extracting food particles either from water through which they move or from a current of water passed over a food-collecting surface. Such a surface frequently is covered in cilia, usually two sets at right angles, one set producing the current and the other trapping food particles, which may be bound in secreted mucus before being conveyed to the mouth. Ciliary feeding occurs, for example, in tubicolous polychaetes, bivalve molluscs and animals with lophophores. Note that 'filter-feeding' is a misleading description of ciliary feeding, because cilia are too small and too closely packed for water to pass between them. Arthropods have no cilia, but cuticular hairs on feeding appendages may trap small particles from water.

Detritus feeding

Ingestion of particles of decomposing organic matter and microbes contained within detritus was probably the ancestral feeding method, and remains an important basis for aquatic food chains.

Large particles or food masses

These are needed by most large animals.

Deposit feeding

The consumption from the substratum of organic detritus is called deposit feeding. Detritus may be swallowed whole into the animal's body, where food is extracted, and (for example in *Arenicola* and earthworms) the familiar worm casts are left behind.

Intake of plant food

Plant food may be obtained by browsing and grazing, or by scraping or boring. Tools are usually required, such as the molluscan radula or the jaws of arthropods and many worms, and chemical secretions may assist food gathering. Hard structures are necessary also to penetrate the cellulose cell walls. Digestion of plant food is difficult for animals, since they lack the enzymes which break down cellulose. Animals frequently retain symbiotic bacteria, fungi or protista in a special gut compartment. The ruminant stomach is a familiar example: many invertebrates that feed on seaweed (for example sea urchins, a few littoral molluscs) or land plants (for example many arthropods and molluscs) similarly rely on microbial symbionts for breakdown of cellulose.

Living animals

Live animals may be the prey either of resident parasites (see Box 6.1) or of carnivores, which kill and eat their prey. Examples of predators occur in all phyla (there are even improbable carnivores among sponges and ascidians). Attack frequently requires poison (e.g. in cnidarian nematocysts, nemertine proboscides, arachnid poison glands) as well as, or instead of, hard structures.

Fluid feeding

Piercing and sucking

Plant or animal bodies may be pierced or sucked by parasites or by other predators.

Uptake of organic food in solution

Soluble organic food may be taken up over any permeable body surface, and this is now known not to be confined to endoparasites but to be widespread in aquatic, particularly marine, organisms. It is usually a secondary development and an auxiliary feeding method, but it can be the main feeding method in some animals without guts.

can be serviced by living cells. External spines justify the name of the phylum ('spiny skins').

17.1.3   The nervous tissue
The nervous system is a diffuse nerve network, condensed in places into nerve fibres, but without a brain. This suggests a degree of coordination no greater than that of cnidarians, yet groups of nerve cells can, as will be explained, become temporary coordinating centres.

17.1.4   Mutable connective tissue
The connective tissue of echinoderms is a most striking unique feature. It is capable of rapid and reversible change in stiffness, often within seconds, under nervous control. The mechanism is a change in the viscosity of collagen fibres in the connective tissue, caused by ionic movements that alter weak interactions between extracellular macromolecules. The process is controlled by calcium binding to neurosecretory cells. The result is that muscle tone can be altered without the expenditure of energy involved in muscular contraction. Examples include the stiffening of sea urchin spines, which can anchor the animal, the stiffening and relaxation of the body walls of starfish and sea cucumbers, the autotomy of arms in brittle stars and the maintenance of tone in the extended arms of sea lilies. It is a property that the human animal can only envy.

17.2   What is unusual but not unique about echinoderms?

In addition to their unique features, echinoderms differ from most other invertebrates in various ways.

17.2.1   Pentaradiate symmetry
Radial symmetry is a legacy of their evolution from sessile animals that collected food from all sides. Symmetry is perhaps not as closely related to phylogeny as once was thought: it is closely related to the feeding methods of the animals concerned. Secondary radial symmetry is familiar in, for instance, many ciliary-feeding annelids and in animals with lophophores. Most present-day echinoderms are motile animals, needing special adaptations of the ancestral radial symmetry. Why echinoderms usually have five rather than any other number of arms is not clear.

17.2.2   Transport of food and removal of waste
Food and metabolic waste may be transported by the water vascular system; amoebocytes in the coelom branches may carry metabolites to any permeable area of skin, where also gaseous exchange occurs. To say that echinoderms have no excretory system is to fall into the

common error of considering only the adult stage in the life cycle: starfish larvae have typical metanephridial excretory systems (see Box 9.1) with ultrafiltration driven by cilia. In adults, there are no nephridia but the axial organ (Figure 17.1e) may assist waste removal.

Osmotic regulation is almost completely absent. The body fluid is not always osmotically identical to sea water – for example, the fluid in the water vascular system of a starfish is slightly hypertonic to the sea – but differences are small. There are no freshwater echinoderms.

17.2.3   Deuterostome development
Cleavage is radial and regulative, without the early determination of cell fate characteristic of spiral cleavage. The blastopore becomes the anus and the mouth opens elsewhere. The mesoderm is derived from the gut wall as the coelomic cavities are formed by a series of out-pouches (see Box 5.2).

17.2.4   The complexity of the coelomic cavities
Echinoderms are unsegmented; the coelom is not divided along a body axis as in an annelid, but it exists as a number of separate divisions persisting from its embryological origins. In addition to the main water vascular system, a starfish has a large perivisceral coelom, and haemal (misnamed), perihaemal and gonadal coeloms (Figure 17.1e).

17.3   How do different echinoderms feed and move?

All animals are constrained by their evolutionary history; in the echinoderms this constraint is particularly obvious, and can be related to many of their unusual features. The ancestors of the phylum were bottom-living sessile animals with massive calcareous skeletons. The mouth faced upward and food was collected from all sides. Most modern echinoderm adults move slowly over the sea bottom, collecting their food into a mouth on the under surface. Figures 17.1 to 17.7 indicate the range of adult form in the six classes of modern echinoderms.

17.3.1   Echinoderm larvae
The larvae feed and move very differently from the adults. Most echinoderms have ciliated filter-feeding larvae swimming in the marine plankton (see below).

17.3.2   Crinoidea
The Crinoids (sea lilies and feather stars) are the oldest surviving group of echinoderms, and resemble their ancestors in being essentially sedentary suspension feeders collecting their food into

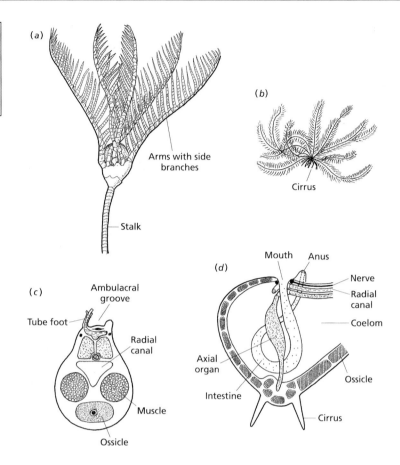

Fig. 17.2 The structure of crinoids: (a) a sessile crinoid; (b) the feather star *Antedon*, a motile crinoid; (c) transverse section of arm, *Antedon*; (d) vertical section of crinoid disc.

an upwardly facing mouth (Figure 17.2a,d). Most of the sea lilies are extinct. In living species the arms and the stem are stiffened by ossicles and mutable connective tissue: both can contract actively in the total absence of muscle. In most species, however, some muscle is present and assists the action after the first quick contraction. Modern feather stars (e.g. *Antedon*, Figure 17.2b,c) hold out their much-branched arms and collect food into the 'ambulacral groove' on the upper surface. Here numerous tube feet bearing mucous papillae act like cilia in conveying the food (bound in mucus) to the central mouth. Feather stars are secondarily motile, being able to swim by waving the arms up and down.

17.3.3 Asteroidea

The asteroids (starfish, Figure 17.1) constitute a large class with some 2000 species. They are predators or scavengers. Some starfish feed by sucking animals into the stomach but others evert the stomach as they feed: typically a starfish grips the two halves of a bivalve shell with the tube-feet suckers and pulls the shell halves apart until it can insert the everted stomach through a tiny gap. The bivalve is then digested inside its shell and sucked into the starfish. Some species are suspension feeders using mucus and cilia.

Walking on tube feet involves alternate attachment to and detachment from the substratum. The soft tube feet can adapt to rough or smooth surfaces, and secretions by two glands in the tube feet give rapidly varying adhesion as the starfish walks; adhesive acidic carbohydrates and proteins alternate with substances of different ionic content that prevent adhesion. Experiments with acidic surfaces have obtained starfish 'footprints'.

Active predators must not only move, they must move in a directed way from place to place. How can the radially symmetrical starfish move in one direction, and without a brain how can it coordinate the stepping of the tube feet in different arms? Observation shows that any arm can become the leading arm, with the other four arms cooperating (Figure 17.3a). Experimental cutting of the nerve ring in two places shows that tube feet in the arms beyond the cuts no longer cooperate with the leading arm, but step towards the tip of their own arm (Figure 17.3b). In an isolated arm, the tube feet step in coordination towards the tip of the arm only if the junction between the radial nerve and the nerve ring is present (Figure 17.3c,d). Further analysis has been possible by direct recording from electrodes in the ophiuroid 'giant neurons', which are several orders of magnitude larger than those of asteroids. The site of coordination is not, as previously thought, in the central

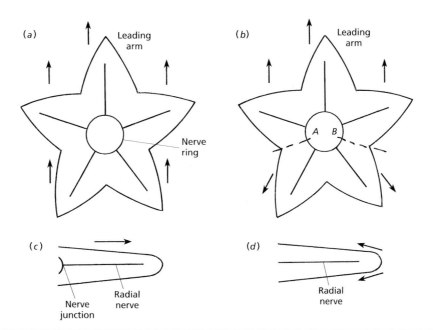

Fig. 17.3 Experimental demonstration of the coordination of tube feet by nerves in an asteroid starfish. Arrows denote the direction of stepping of tube feet in (a) an intact starfish; (b) a starfish with the nerve ring cut at A and B; (c) an isolated arm including the junction between the nerve ring and the radial nerve; (d) an isolated arm without the junction.

ring but in the radial nerve cord: patterns of activity in any one arm can be conducted round the ring and direct the stepping of tube feet in the animal as a whole. In this way a starfish can make a temporary 'brain' without having a permanently defined brain structure.

17.3.4 Ophiuroidea

The ophiuroids (brittle stars) constitute another large class (2000 species) of rather smaller starfish. They have very much more hard skeleton than asteroids and, although they are stellate in form, the arms are clearly marked off from the disc (Figure 17.4a). Most of the arm consists of central muscle surrounded by ossicles (Figure 17.4b,c) which are able to move on each other and give the arms great flexibility (hence the name 'serpent stars'). These ossicles are often called 'vertebrae', and indeed no other invertebrates have such vertebrate-like structures. Most ophiuroids resemble asteroids in being active carnivores, but their methods of feeding and locomotion are very different. Food is seized by the mobile arms, assisted by mucus secreted by tube feet that have no ampullae and play only a minor role. Ophiuroids walk by two arms on each side sweeping back from the leading arm; this 'rows' the animal along in a series of jerks, with the spines providing traction against the ground. Swimming is achieved by similar arm movements. Coordination of these relatively rapid movements is more highly developed than

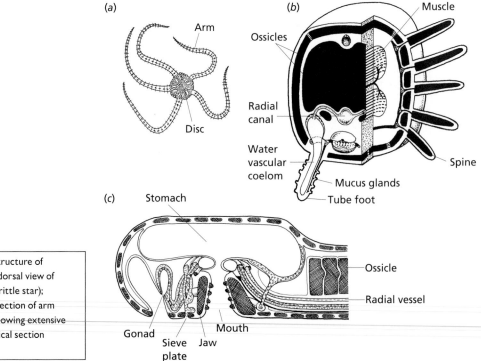

Fig. 17.4 The structure of ophiuroids: (a) dorsal view of an ophiuroid (brittle star); (b) transverse section of arm of *Ophiothrix*; showing extensive ossicle; (c) vertical section of ophiuroid.

in asteroids. The arms not only have 'segmental' ossicles, they have a corresponding series of nerve ganglia along each arm. As described above, dominance by the leading arm appears to depend on the action of the arm's nervous tissue as a whole.

Some species have marked sensitivity to light intensity, enabling them to detect shadows of predators, change colour and escape rapidly. Photosensitivity in other echinoderms is confined to diffuse dermal receptors, but light-sensitive ophiuroid species incorporate calcite crystals into the skeleton and arrange them into microstructures acting as double lenses, each especially sensitive to light from a particular direction. This is a nice example not only of the more elaborate sensory structures required by faster-moving invertebrates (compare Cubozoa, Chapter 4) but also of a structure combining mechanical and sensory functions.

17.3.5 Echinoidea

The echinoids (some 950 species) include 'regular' echinoids (sea urchins, Figure 17.5a,b) and 'irregular' ones (heart urchins, sand dollars and others, Figure 17.5c). Unlike starfish and brittle stars, sea urchins are browsing herbivores and very slow-moving. Their speed varies partly according to their habitat: a larger sea urchin moves faster on a horizontal surface, but uses more energy in moving on a vertical one. Echinoids are spherical or egg-shaped, with the axis between the oral and aboral surfaces much extended (Figure 17.6). Uniquely among echinoderms, the body is encased in a fixed lattice of ossicles, which may be fused to form a rigid 'test'. The body is covered in spines, and pedicellariae are important in removing

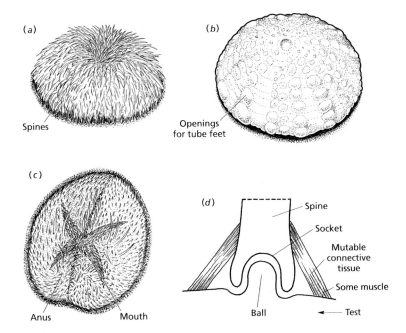

Fig. 17.5 The structure of echinoids: (a) external view of a sea urchin (echinoid); (b) a sea urchin 'test' (skeleton); (c) a sand dollar, an irregular echinoid; (d) the mobile spine of a sea urchin.

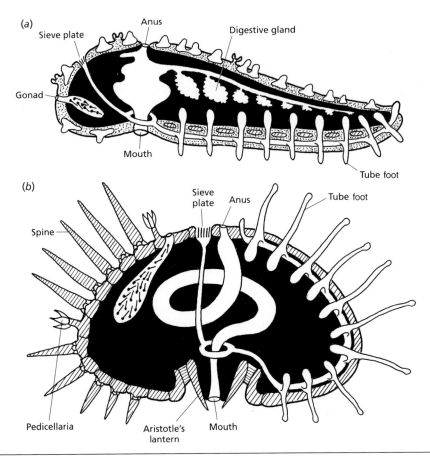

(a)

Sieve plate

Anus

Digestive gland

Gonad

Mouth

Tube foot

(b)

Sieve plate

Anus

Tube foot

Spine

Pedicellaria

Aristotle's lantern

Mouth

Fig. 17.6 Diagram comparing the general form in vertical section of (a) a starfish and (b) a sea urchin.

settling organisms. In place of arms there are five ambulacral grooves, where tube feet emerge between small skeletal plates (Figure 17.5b). The rigid test means that water influx from the sea, while still essential, is not so immediately related to tube foot extension, because some water is retained in the animal.

Sea urchins feed by scraping seaweed from rocks with an elaborate structure of movable skeletal plates called 'Aristotle's lantern'. This consists of five jaws and one 'tooth', connected to the jaw by a central ligament in which the connective tissue is made hard or soft, as the food-supporting surface may require. Locomotion is mainly by spines that articulate with the main skeleton by ball-and-socket joints (Figure 17.5d) and are controlled partly by muscle and partly by mutable connective tissue. The tube feet are very well developed, with ampullae and suckers. They assist locomotion and keep the urchins close to the substratum on which they are grazing or anchor them in crevices.

Sea urchin gonads are sometimes regarded by humans as a delicacy: as is possible in a slow-moving animal, they contain a reserve of stored food. When the urchin encounters a particularly rich food source, there is a rapid increase in gonadal tissue.

Irregular echinoids are deposit feeders adapted to burrowing shallowly under sand. Some are greatly flattened and covered in small spines, with bilateral symmetry superimposed on the radial plan. The tube feet are very diverse in structure and function, some being modified for digging and others as respiratory 'siphons' leading up to the surface of the substratum.

17.3.6 Holothuroidea

The holothuroids (sea cucumbers) comprise about 900 species which are very unlike other echinoderms. They are secondarily bilaterally symmetrical, lying on one side with an elongated body axis between the mouth and the anus (Figure 17.7a,b). The endoskeleton is very much reduced, leaving a muscular body wall with a few embedded ossicles but without spines or pedicellariae. There is only one gonad. Respiratory trees, branches of the gut taking in oxygenated sea water, are characteristic.

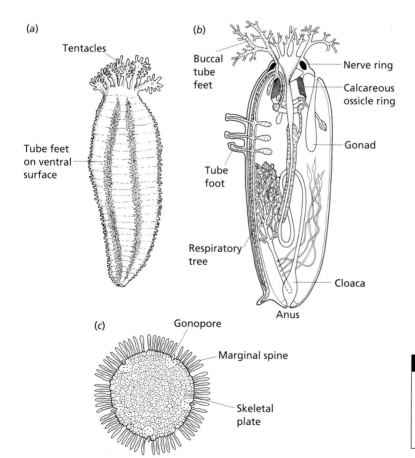

(a) Tentacles
Tube feet on ventral surface

(b) Buccal tube feet
Nerve ring
Calcareous ossicle ring
Gonad
Tube foot
Respiratory tree
Cloaca
Anus

(c) Gonopore
Marginal spine
Skeletal plate

Fig. 17.7 The anatomy of holothuroids and concentricycloids: (a) diagram of a holothuroid; (b) vertical section of a holothuroid; (c) dorsal view of *Xyloplax medusiformes*, a concentricycloid.

Typically holothuroids are deposit feeders. A group of anterior tube feet forms long buccal tentacles that are held out as a net or gather food from the sand. Locomotion is slow, by worm-like wriggling of the muscular body wall, or there may be locomotor tube feet in ambulacral grooves on the underside of the body. When pursued by predators, holothuroids may immobilise them by extruding a mass of very sticky blind-ending tubules, or may break off the internal organs and leave them in the path of the predator: this is achieved by complete slackening of the ground substance of the mutable connective tissue. The rest of the holothuroid escapes, and regenerates its viscera.

17.3.7 Concentricycloidea

The concentricycloids (sea daisies) were discovered only in 1986 in deep seas off New Zealand; a second species of *Xyloplax* has now been found in the Caribbean. Concentricycloidea are at present considered to constitute a separate class closely related to Asteroidea: they may even be placed within that class. Superficially they resemble medusae (Figure 17.7c) but there is a water vascular system with two connected concentric rings, one of which bears tube feet. The spermatozoa are very unusual and, uniquely in the phylum, fertilisation may be internal. Feeding methods are unknown (one of the species has no gut) but absorption of dissolved organic materials is probably important.

17.4 Do the larvae illuminate echinoderm evolution?

Free-living larvae, in most phyla, are simpler than the adults. This does not mean that they represent ancestral forms; they are evolutionary novelties, adapted to their own way of life. Reference to ancestral forms is especially tempting in those echinoderms which have bilaterally symmetrical larvae – but bilateral symmetry is to be expected in motile animals. Motile planktonic larvae are found in most echinoderms, and are believed to be primitive to the phylum. Some of the species that instead develop directly from yolky eggs are found to contain vestigial larval structures, which supports the idea that direct development is a secondary specialisation. Planktonic larvae are an advantage to sessile or slow-moving animals living on the sea bottom, because they can exploit a different food source and can disperse the species. They also allow greater fecundity, since a large number of small eggs without food reserves can be produced.

Figure 17.8 indicates the larvae characteristic of the different classes. The prevalence of convergent evolution is instantly apparent. Larvae are above all adapted to their own way of life. That both ophiuroids and echinoids have a 'pluteus' larva with the ciliary band extended into arms by eight skeletal rods does not mean that brittle

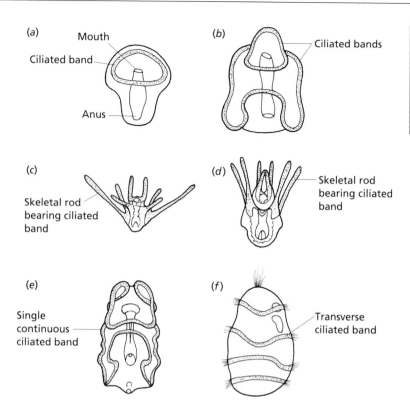

(a)

Mouth

Ciliated band

Anus

(b)

Ciliated bands

(c)

Skeletal rod
bearing ciliated
band

(d)

Skeletal rod
bearing ciliated
band

(e)

Single
continuous
ciliated band

(f)

Transverse
ciliated band

Fig. 17.8 Echinoderm larvae:
(a) dipleura, the first larva of
most groups; (b) asteroid
bipinnaria; (c) ophiuroid
ophiopluteus; (d) echinoid
echinopluteus; (e) holothuroid
auricularia; (f) crinoid doliolaria.

stars and sea urchins have an improbably close relationship, but
that both classes have the same adaptation, which extends their
surface area thereby both facilitating floating by increasing the
viscous drag and increasing the feeding surface (some echinoid larvae
grow longer arms when food is short). Some ophiuroids have larvae
similar to the crinoid doliolaria, derived by loss of the arm rods,
but this is no link between ophiuroids and crinoids. Nor has the
resemblance between the asteroid bipinnaria and the holothuroid
auricularia any phylogenetic significance except that both are
developments of the basic 'dipleurula' larva (Figure 17.8a) in which
form many echinoderms hatch. Sea lilies were thought to be an
exception, but a dipleurula larva has recently been found, in very
deep water.

Nearly all of these larvae are bilaterally symmetrical; the adult
rudiment is set aside in the late larva and radial symmetry first
appears when it develops. Convergent evolution extends much
further than the external form of different larvae. The loss of
planktonic larvae from the life cycle has occurred many times in
unrelated groups, and the same series of steps can be traced: first
the eggs become larger, then the larvae lose the ability to feed, and
then the time before metamorphosis is reduced. A feeding larva has
been lost in many ophiuroids and 14 times in five different orders
of echinoids. Total loss of larvae has occurred in several ophiuroids
and in at least six echinoid lineages. These changes may or may not

be accompanied by smaller adult size, viviparity and brooding behaviour. In asteroids also, simplification and reduction of larval form has occurred many times separately, as has the change to hermaphroditism or to viviparity. Interesting recent work has shown that (as in tadpoles) metamorphosis can be accelerated by the thyroid hormone, thyroxine, which may be internally produced or supplied externally. It is now thought that thyroxine may be important in regulating the life histories of echinoderms in general.

Larval morphology clearly has diversified independently of adult morphology, and convergent adaptations are rife. Can larval structure illuminate phylogenetic relationships at all? Probably only in those characters that are the immediate consequences of early development being deuterostome: for example the mouth is surrounded by cilia, which become the single feeding band of the earliest larva, retaining food particles on an upstream beat. This is significantly different from the trochophore larva of protostomes, where the ciliary bands are differently arranged and trap food particles as they travel downstream. The larvae of other deuterostomes, the hemichordates and chordates, capture their food in the same way as echinoderm larvae.

The conclusion is that we can learn about the process of evolution from echinoderm larvae (the selective advantage of having a larval stage in the life cycle and the adaptations which have occurred) but, owing to the abundance of convergence in larval forms, we can learn very little about the pattern of evolution. Chapter 20 explains how we can trace the course of echinoderm evolutionary history from the particularly good fossil record for many classes of echinoderms, and how molecular methods (including analysis of the genes controlling early development) can help us further.

Box 17.1 | Larvae

What is a larva?

A larva is a pre-adult form in which some animals hatch from the egg. Further definition runs into difficulties: a larva is different in form from the adult of the species, but the difference may be confined to smaller size and immaturity, as in the nematodes and exopterygote insects, where the series of larvae simply represent a discontinuity imposed on growth by the need to moult. A larva is sexually immature (but see the discussion of paedomorphosis in Chapter 18). Characteristically it is a feeding stage in the life cycle, but many larvae do not feed (e.g. the short-lived cypris larva of barnacles and the tunicate tadpoles, both of which serve simply to disperse the sessile adults). Larvae often are dispersal agents (as in many crustaceans) but not for example in insects, where the adult may have wings.

Why have a larva?

Early hatching can be an economy if it saves the adult from providing large quantities of yolk. Larger numbers of young can be produced and the reproductive turnover will be faster, but most larvae are fragile and very many will be eaten or die. Larvae often exploit a food source different from the adult's, and may colonise a different habitat, thus avoiding competition between adult and young. They may both disperse the species and find a suitable habitat for the adult.

Which invertebrates have larvae?

All phyla have examples of life histories including larvae (see figures in the relevant chapters). In general:

Marine invertebrates. The marine plankton is rich in food, readily disperses animals, and small floating forms do not require special adaptations. In annelids and crustaceans, for example, marine animals nearly all have larvae, while freshwater and terrestrial animals do not.

Sessile animals. Slow-moving bottom-living animals almost always use larvae for dispersal. Examples include sponges, hydroids, anemones and corals, marine molluscs other than cephalopods, barnacles and echinoderms.

Parasites. Larvae are important agents of transfer between parasite hosts. Often there is an elaborate life cycle with a succession of hosts, commonly with one preying upon another, and the volume of reproduction may be increased by asexual reproduction of some larval stages. Figure 6.4 gives examples of two platyhelminth life cycles.

Arthropods need to moult as they grow and therefore readily incorporate metamorphosis in the life cycle. Larvae can be very unlike the adults.

What can larvae tell us about phylogeny?

The answer has to be 'very little'. Larvae are adapted to their environment. However, larvae are the products of early hatching, and therefore may show in their structure the consequences of earlier embryonic development, such as the method of cleavage and of mesoderm formation, which do have phylogenetic significance. Many protostomes, for example, hatch as larvae rather like trochophores: this does not necessarily relate them closely to annelids or to molluscs, but it does distinguish them from echinoderms and chordates. Occasionally aberrant forms (such as *Sacculina*: see Chapter 13) can be placed by larval resemblances, but more often such resemblances must be attributed to convergence. Whether larvae are primitive, to invertebrates as a whole or in particular phyla, is uncertain; this is discussed for echinoderms in Chapter 17.

This brief survey of larvae is very far from complete, even concerning invertebrates mentioned in this book, omitting for example the sessile scyphistoma larva of many jellyfish (Chapter 4) and the great range of aquatic and terrestrial larvae of insects (Chapter 15), but it provides an introductory overview.

Chapter 18

Invertebrate Chordata and Hemichordata

The phylum Chordata consists mainly of the class Vertebrata (animals with backbones, i.e. the fish, amphibians, reptiles, birds and mammals, including ourselves). There are, however, some smaller groups of animals that do not have backbones but possess the distinctive chordate characters. All are marine, and like the echinoderms they are deuterostomes. These animals are interesting because they demonstrate other varieties of invertebrate life and because they provide information about deuterostome relationships and the origin of vertebrates.

Hemichordates are not chordates. They were unfortunately named before their structure was fully understood and subjected to molecular investigation. They are now recognised as a separate phylum among the deuterostomes; as will be explained, they are closer to echinoderms than to chordates.

CHORDATA

18.1   What are the chordate characters?

Figure 18.1 shows the distinctive characters: the notochord, the hollow dorsal nerve cord, the pharyngeal gill slits and the post-anal tail. All these characters are related to swimming by lateral undulations of a body like a tadpole or a fish.

18.1.1   The notochord

The notochord gives the phylum its name (avoid verbal confusion with 'nerve cord'). It is a dorsal stiffening rod running from near the anterior end to the posterior tip of the body. It is made of close-fitting cells with hydrostatic pressure due to intracellular vacuoles, enclosed in a fibrous sheath; it arises from the roof of the embryonic gut. It is longitudinally incompressible but laterally flexible, and serves as an energy-saving device, because any animal swimming

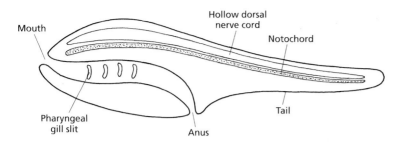

Fig. 18.1 Diagram to show chordate characters.

by muscular undulation must be stiffened to localise the effects of muscle contraction. A swimming leech, for example, uses energy in contracting strong dorsoventral muscles while it generates the muscular wave that pushes it through the water, while a chordate is sufficiently stiffened by the notochord.

18.1.2 The hollow dorsal nerve cord

Immediately dorsal to the notochord, the nerve cord innervates the main swimming muscles of chordates that lack legs. In many other animals, notably annelids and arthropods, the main nerve cord is solid and ventral in position. In chordates the principal blood vessels are ventral and the blood flows forward ventrally, unlike annelids and arthropods, where the blood flows forward dorsally. These differences long ago prompted the suggestion that chordates evolved from 'worms turned upside down'; recent genetic analysis has rescued this suggestion from ridicule and considered it seriously (Chapter 20).

18.1.3 Pharyngeal (gill) slits

Pharyngeal slits primitively served filter-feeding, from a stream of water due either to the animal swimming through the sea or to ciliary beating bringing sea water through the animal. Water enters by the mouth, folds of skin (gills) aided by mucus trap the food particles, and the water leaves the pharynx through the gill slits instead of passing right through the gut to the anus. Chordates feeding on larger masses of food may use the gills for respiratory exchange and retain the gill slits to let water out.

18.1.4 The post-anal tail

Like the notochord, the tail is concerned with efficient locomotion, being primitively a muscle-packed region exerting propulsive force.

These four characters may be lost or modified in many members of the phylum. For example in most vertebrates the embryonic notochord is replaced during development by the vertebral column (backbone), and in ourselves the gill slits are only briefly marked in the early embryo and the tail is vestigial. Similarly, many invertebrate chordates only show the chordate characters at certain stages of the life cycle.

18.2 Which are the invertebrate chordates?

18.2.1 Cephalochordata

The cephalochordates (25 species, up to 100 mm in length) show the chordate characters very clearly. The lancelet or 'amphioxus', *Branchiostoma*, is the main example (Figure 18.2). The amphioxus tapers to a point at both ends; the simple epidermis extends into dorsal, caudal and ventral fins, and by transparency the V-shaped body muscles are very obvious from outside. It can swim by lateral undulations but spends most of its time half-buried in shallow marine gravel: it can still be found by dredging near the Eddystone lighthouse in Plymouth Sound, England. The amphioxus is specialised in that the muscles and their nerves are asymmetrical in arrangement and the gills are greatly elaborated and multiplied, with accessory ciliary tracts and an external oral wheel. Food particles are trapped in mucus secreted by a structure called the endostyle on the floor of the pharynx. The gill slits are very numerous and do not open directly to the outside world (so many perforations would greatly weaken the body wall) but into an atrium, which opens more posteriorly by an atriopore. The atrium is formed by folds of the body wall in the pharyngeal region, which appear in the late larva and grow down to the ventral surface, where they fuse (Figure 18.2a,b). The notochord is unique in extending right to the anterior tip of the body. The anterior end of the dorsal nerve cord is not expanded into a brain, nor is there a heart; the blood is circulated by general contractility of the blood vessels. The excretory organs are small and not well developed; they are intermediate between protonephridia and metanephridia in structure. They are hard to detect with only light microscopy and earlier descriptions are

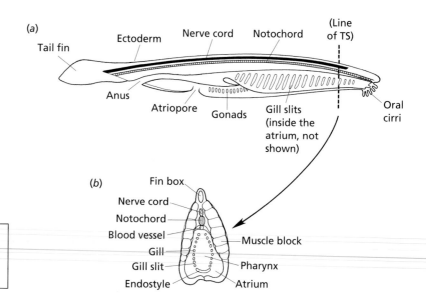

Fig. 18.2 A cephalochordate, the amphioxus (*Branchiostoma*): (a) longitudinal section; (b) transverse section (TS).

misleading. The sexes are separate, fertilisation is external and the zygote hatches early as a small free-swimming version of the adult.

18.2.2 Urochordata (tunicata)

These are our most improbable cousins, revealed as such only by their larvae. This 'tadpole larva' (Figure 18.3a) shows all the chordate characters, but at metamorphosis it turns into a sea squirt or a colony of similar organisms, enclosed in a gelatinous tunic.

> **Ascidiacea** (ascidians, *c.* 2000 species, 1 mm–1 metre in length): sessile filter-feeding 'sea squirts', common in shallow sea water. There is no head and the nervous system is poorly developed. The gut is U-shaped between two openings, the buccal and atrial siphons, and the much enlarged pharynx is perforated by many gill slits; this is the only chordate character remaining in the adult (Figure 18.3b). As in the amphioxus, ciliary feeding is assisted by mucus secreted by an endostyle, and the gill slits

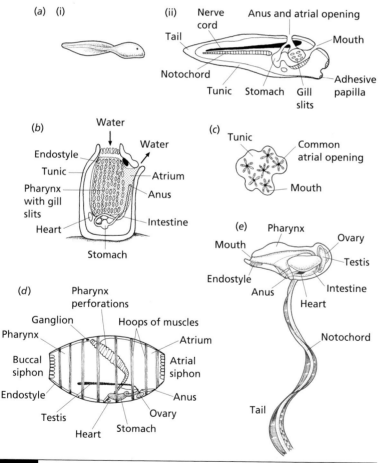

Fig. 18.3 Urochordates: (a) tadpole larva of a tunicate, (i) external view, (ii) longitudinal section; (b) vertical section of *Ciona*, an ascidian; (c) *Botryllus*, a colonial ascidian; (d) *Salpa*, a thaliacean; (e) *Oikopleura*, a larvacean.

open into an atrium. Unique features of ascidians include a heart that periodically reverses its beat and the enclosure of the whole body in a gelatinous tunic which is strengthened by cellulose fibres and penetrated by blood vessels. The tunic is not moulted but grows with the animal. Another unique feature is that as sea water is passed through the animal, vanadium ions are retained and concentrated, mainly in the blood cells. Apparently the vanadium ions assist the polymerisation of fibres in the tunic, and may also act as antibiotics.

Ascidians are nearly all hermaphrodites, usually cross-fertilised, and the zygote hatches as the tadpole larva, which is primarily an agent of dispersal. Great powers of regeneration and budding are characteristic of ascidians, as is manifested by compound ascidians. Individuals are combined in a series of star-shaped colonies, very common on the seashore and often brightly coloured, embedded in jelly. Each colony has separate mouths at the points of the star and a common atrial opening in the centre (Figure 18.3c).

Thaliacea ('salps', 70 species): floating transparent tunicate colonies (Figure 18.3d), with buccal and atrial sinuses at opposite ends of the colony. The water current, generated by muscular contraction of the body wall, is used for swimming by jet propulsion as well as for filter-feeding and respiration. Salps are widely distributed and are now becoming more common in the southern oceans, as they can withstand rising temperatures better than krill. Sadly salps are no substitutes for krill as a basis of food chains.

Larvacea (70 species), best known from *Oikopleura* (Figure 18.3e). They are more obviously chordate than other adult tunicates since the larval form is retained following accelerated early development. The adult floats in the plankton enclosed in a gelatinous 'house' (without cellulose fibres) through which tail movements propel a stream of water. There is only one pair of gill slits, and the food consists of very small food particles trapped in mucus. The house is shed and reassembled at frequent intervals.

Sorberacea. These deep-water forms comprise a few species of most remarkable carnivorous tunicates. Prehensile finger-like extensions of the buccal siphons stretch out and capture small worm and crustacean prey.

18.3 How are the invertebrate chordates related?

Cephalochordates appear to be closely related to craniates (vertebrates). Morphology and embryonic development apparently united these groups beyond reasonable doubt; the only question has been whether the amphioxus is primitively simple or has secondarily lost the vertebrate specialisations. Genetic evidence confirms the primitive nature of the amphioxus but questions the closeness of the relationship (see Chapter 20).

Urochordates have been very hard to place. A classic early account by W. Garstang, followed by N. J. Berrill, invoked 'neoteny', where the larval form becomes sexually mature; if a tadpole larva were to reproduce, the result would be an animal very like the amphioxus (or at least, very like its larva before adult specialisations develop). The sessile and gelatinous tunicate adult would then be entirely omitted from the life cycle. Species showing both kinds of life cycle (with an adult in the sessile tunicate or the chordate form) survive to the present day, placed in different phyla.

The phenomenon of 'paedomorphosis' (reproduction in the larval form) is known from a number of present-day animals. It is called 'progenesis' when gonadal development is accelerated and reproduction occurs earlier in the life cycle, or 'neoteny' when somatic development of adult tissues is delayed and reproduction occurs at the time normal to that species but while it is still a larva. Paedomorphosis is one way in which the normal process of natural selection can result in sudden abrupt changes in form. It might have occurred several times in the chordate lineages: larvaceans such as *Oikopleura* might be products of neoteny within the tunicates, and the vertebrates resemble the larva rather than the specialised adult. Chapter 20 discusses the role of neoteny.

Because of the great interest in the origin of vertebrates, the full genome of the ascidian *Ciona intestinalis* has recently been sequenced (see Chapter 20), but it is not possible to trace the course of chordate evolution with certainty. What is certain is that the chordate characters, at whatever point they arose, proved to be excellent adaptations to a free-swimming life in the sea.

HEMICHORDATA

18.4 What are the hemichordates?

They consist of two groups, very dissimilar in appearance.

18.4.1 Enteropneusta

The enteropneusts (70 species, 9–200 cm in length) are large solitary worm-like animals such as *Balanoglossus*, burrowing in the mud or sand of the shallow seas. The body consists of three regions, a proboscis, collar and trunk (Figure 18.4a,b), corresponding to the three divisions of the embryonic coelom, the protocoel (proboscis), the mesocoel (collar) and the metacoel (trunk). The ciliated proboscis collects food into the mouth, which opens from the collar. The lobe in front of the mouth has a complex 'heart/glomerulus' system,

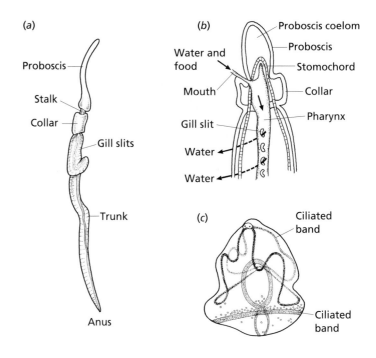

Fig. 18.4 Hemichordates:
(a) *Balanoglossus* (= *Saccoglossus*), a solitary enteropneust;
(b) diagrammatic longitudinal section of the anterior end of an enteropneust; (c) tornaria larva of an enteropneust.

(a)

Proboscis

Stalk

Collar

Gill slits

Trunk

Anus

(b)

Water and food

Mouth

Gill slit

Water

Water

Proboscis coelom

Proboscis

Stomochord

Collar

Pharynx

(c)

Ciliated band

Ciliated band

supported by a rod-like 'stomochord'; the protocoel opens by a pore and the paired metacoels have ducts to the exterior by way of the first gill slits. The trunk is perforated by a long row of gill slits, assumed to serve respiration: there is no filter-feeding (food is collected by the proboscis) and there is no mucus-secreting endostyle. The nervous system is not well centralised: there is a nerve net resembling that of echinoderms, epidermal in origin and position. The main concentration of neural tissue in the collar is hollow, and develops much as in chordates.

The larva is not a tadpole but a 'tornaria', remarkably similar to the holothuroid auricularia in appearance (Figure 18.4c, compare Figure 17.8e) and to the bipinnaria of various echinoderms in structure. In both the tornaria and the bipinnaria there is a short pore canal leading from the most anterior cavity to a 'hydropore', functioning as an excretory outlet. The tornaria also resembles most echinoderm larvae in having a gel-filled earliest body cavity, permitting development of a large larva with little cellular material.

18.4.2 Pterobranchiata

The pterobranchs (21 species in two genera, *Cephalodiscus* and *Rhabdopleura*) are minute (1–5 mm long) sessile and colonial animals, ciliated all over, with lophophores bearing tentacles that collect their food (Figures 18.5a and 16.2a); these lophophores are remarkably similar to those of the three protostome phyla discussed in Chapter 16. Pterobranchs are undoubted deuterostomes, and this resemblance must be due to convergence.

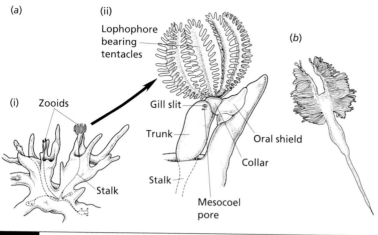

Fig. 18.5 Hemichordates: (a) *Cephalodiscus* colony, with one zooid enlarged; (b) a fossil graptolite.

Cephalodiscus has a single pair of gill slits; *Rhabdopleura* has none. There are three divisions of the coelom: the protocoel in the lophophore, the mesocoel, having a pore opening to the outside, and the metacoel. There is a cephalic shield, a glandular and ciliated disc over which zooids glide inside their tubes. The very simple nervous system is entirely epidermal. Pterobranchs reproduce asexually by budding or sexually by releasing gametes. The larva, unlike the enteropneust tornaria, is uniformly ciliated and short-lived. It has a store of yolk and does not feed, serving solely for dispersal of these sedentary animals.

Pterobranchs can be traced back in the fossil record to the graptolites (Figure 18.5b), known from the Cambrian to the early Devonian; they were once considered to be cnidarians, but further study diagnosed them as pterobranchs. There was great excitement when in 1989 a living graptolite was hauled up from the deep sea and could (although this is controversial) be assigned to a modern genus, as *Cephalodiscus graptolitoides*.

18.5 What do enteropneusts and pterobranchs have in common?

They appear to be so different, but the same basic structure can be traced and they have several unique features in common:

The stomochord is a rod-like hollow process growing out from the gut dorsally, lined with flagellated cells. It projects forward from the roof of the mouth into the proboscis in enteropneusts (Figure 18.4b), where it supports the heart and excretory organ, or to the oral shield in pterobranchs (Figure 18.5a). This rod looks like a short notochord and was once described as such, but this was an error: it develops differently from a notochord, and is a glandular organ not a supporting rod. It is associated with the excretory organ in both groups.

The collar is a distinctive structure containing the mesocoel, with ducts taking coelomic fluid to the first gill slit. In both groups it bears the mouth on the ventral side and is the main site of the epidermal nervous system.

The 'heart/glomerulus' complex is similar in both groups. The heart is situated in the posterior part of the protocoel and receives blood from a channel between the stomochord and the pericardium. Blood is driven into the vessels by the contraction of the pericardium. Folds of the closely associated gut wall form the glomerulus, the site of filtration. Urine is expelled through the protocoel pore.

The deep sea floor provides intriguing animals that at first were thought to combine features of both groups (they were called 'lophoenteropneusts'), but further investigation showed them not to have tentacles but to be a new family of typical enteropneusts.

18.6 Where do the hemichordates fit in to the deuterostomes?

18.6.1 Resemblances to chordates

Resemblances to chordates are few. There is no notochord, no tail, no ventrally forward blood flow and although the nervous tissue in enteropneusts is hollow it is not extended into a chordate dorsal nerve cord. Larval structures do not provide evidence of any relationship to chordates. Unlike chordates, both hemichordate groups collect their food externally, enteropneusts by the proboscis and pterobranchs by the lophophores. The gill slits are the outstanding 'chordate character' of enteropneusts, and are presumably necessary to assist respiration in these burrowing animals.

At this point a wider view of deuterostome relationships is required, to include the echinoderms, the only other major deuterostome phylum. Recently, the idea that gill slits are confined to chordates has been questioned, most dramatically by the finding of gill slits in at least one group of early echinoderm fossils. Could gill slits be primitive deuterostome characters, secondarily lost in echinoderms?

18.6.2 Resemblances to echinoderms

Despite the very different adult structure, developmental characters suggest a close relationship between echinoderms and hemichordates. Their early development is very similar, not only in the earliest stages that define deuterostome embryology but also in the origin and development of the three divisions of the coelom. For example the water vascular system in echinoderms is derived from the mesocoel, as is the collar coelom which penetrates throughout the lophophore in pterobranchs. The sessile crinoids collect their food using tube feet containing a branch of the water vascular system, very much as the sessile pterobranchs use the lophophore. The echinoderm madreporite corresponds to the hemichordate collar pore. In both phyla the left coelomic pouches become dominant as the animal develops, and the pouches on the right side are lost.

The tornaria larva of enteropneusts has resemblances to the echinoderm bipinnaria that are more than superficial. Both larvae have a pore canal from the anterior coelom leading to a hydropore, now known to function as an excretory pore. In both groups this larval structure is carried over into the adult. Chapter 20 shows how these deductions from the morphology of these curious animals are supported by genetic evidence.

Box 18.1 | Deep-sea invertebrates

The deep sea

Many invertebrates occur at great depths, often with unexpected characteristics. Examples already given in this book include hexactinellids and carnivorous sponges (Chapter 3), jellyfish (Chapter 4) and ctenophores (Chapter 5), annelids (Chapter 9), the mollusc *Neopilina* (Chapter 10), septibranchs, *Nautilus* and giant squids up to 15 metres in mantle length (Chapter 11), copepods, shrimps and amphipods (Chapter 13) and carnivorous tunicates (Chapter 18).

Deep-sea invertebrates are zoned according to the depths they can tolerate. In the twilight zone (c. 150 m down) there are many large invertebrates, often with very large eyes and transparent bodies that may escape detection. Below 200 m there is no photosynthesis; below 1000 m there is no light at all. Problems at such depths include:

Food and oxygen shortage. The food chain is based on microorganisms and on organic material dropping from surface water, or dissolved in the sea and absorbed directly. In deeper water energy comes from geochemical sources, largely hydrogen sulphide (H_2S) broken down by chemosynthetic bacteria. Here predatory animals are often luminescent, to communicate and to attract potential food. For the prey, dark red colour is protective at dark levels and vertical migration is common: up at night, since shallower water contains more food; down by day, to depths with fewer predators. The shortage of food and oxygen means that few deep-sea animals can have active lifestyles.

Pressure increases by about 10 atmospheres per 100 m depth, giving 100 atmospheres at 1000 m. Special adaptations are required. For example, bodies full of water (as in jellyfish) are cushioned, but water must be removed from proteins to enable them to fold normally. Trimethylamine oxide (TMAO, well known in fish) is present in many invertebrates in quantities increasing with external pressure.

Temperature is less than 5 °C below 1000 m deep.

Hydrothermal vents

Hydrothermal deep-sea vents, mostly chimneys called 'black smokers', were discovered in 1977. They are formed by underwater volcanos making a new sea floor out of hot rocks rich in minerals, and superheated water presses up to the surface and spurts out, making vents with openings about 10 cm in diameter. The hot fluid is often very acidic and has a high concentration of H_2S, methane and metals, especially iron. The temperature is usually about 65 °C, but 390 °C has been recorded. These vents have been found to be common in the Pacific, Atlantic, Indian and Arctic oceans.

Amazingly, the edges of these smokers are inhabited by a large variety of invertebrates: none of these is a completely new kind of animal but most are unfamiliar species of recognisable groups. Over 500 new species have already been found and 12 new families; 90% of the species are endemic (unique to this habitat). They include gastropods (one species covers its foot in scales made from iron salts), bivalves, shrimps gathering near the vents, crabs feeding on bacteria, nemertines with symbiotic sulphur bacteria in their epidermal cells and above all polychaete annelids, especially the vestimentiferan 'Pogonophora' (now called Siboglinidae). Some 200 000 of these were found together 2500 metres down in a black smoker in the Pacific, off Costa Rica, all red or white in colour. *Riftia* is the most striking and characteristic vestimentiferan, sticking up at least 2 m; the tube grows in length by 85 cm in a year, making these the fastest-growing invertebrates known. *Riftia* produces yolky eggs that float away and hatch into yolky larvae; there is never a mouth or a gut at any age.

Animals around vents are zoned horizontally, according to their toleration of high temperatures. The Alvinellidae, a new polychaete family related to the Terebellidae (Figure 9.6c), are common on the sides of black smokers, being able to survive sustained temperatures of 50 °C at the pressure prevailing: high pressure assists survival at high temperatures. All the vent animals depend on chemosynthetic bacteria, by filtering them, farming them and grazing, or by importing them as symbionts. They all have many special adaptations, of membrane and DNA structure and blood and enzyme function, enabling them to survive at high temperatures.

Cold seeps

Cold seeps also occur at similar depths, with surprisingly similar fauna, also dependent on bacteria metabolising sulphur and methane. Here temperatures are low, life is long and growth is slow (a bivalve takes 170–250 years to reach 3 m in length).

The animals in hydrothermal vents and cold seeps are not, as at first thought, 'living fossils', sheltered survivors from earlier epochs. Molecular evidence suggests that most of these animals arose relatively recently. Continuing exploration of these habitats is revealing more and more remarkable invertebrates.

Chapter 19

Development

Development of an individual organism (ontogeny) begins with a single cell, most commonly a fertilised egg. To make a multicellular organism this cell must repeatedly divide, and the daughter cells must grow and become different from each other. Cells must move as they become organised into a body. Development depends upon the coordinated behaviour of cells, and must produce the right sort of animal every time. The main processes of earliest development (briefly introduced in Box 5.2) are discussed in this chapter, using many different examples (occasionally even vertebrates). To control the diversity of development and to bring out general features, four main examples are used: the mollusc (freshwater snail) *Lymnaea*, the echinoderm (sea urchin) *Echinus*, the nematode *Caenorhabditis* (*C. elegans*) and the insect (fruit fly) *Drosophila*. Development is directly or indirectly under the control of genes at every step: our growing understanding of this control is introduced at the end of the chapter.

19.1 How do animals develop?

Cell division: cleavage, the earliest divisions of the fertilised egg, proceeds by DNA replication and subdivision of the cytoplasm, normally without any increase in size (growth occurs later).

Cell differentiation is preceded by **commitment** (sometimes called 'determination'), which is the limitation of developmental capacity, directing the cell to a particular pathway. This may happen long before there are any visible signs of differences between cells, and the timing of commitment in relation to cleavage varies in different kinds of animal.

Gastrulation follows cleavage. It is a series of **cell movements** bringing the cells that will form internal organs to the inside of the developing embryo.

Pattern formation: the organism has to be given a pattern as a whole, and the first step is the establishment of axial polarity and the interactions of cells that give them positional information.

Growth: at each stage in the laying down of the basic pattern of the organism, there can be some growth. Cells increase in size and multiply, and extracellular materials (including fluids and hard skeletons) are produced.

All these processes achieve an ordered increase in biological complexity as the three-dimensional cell architecture of the egg is divided into lineages of developing cells.

19.2   What makes different animals develop differently?

Different invertebrate phyla often have characteristic developmental mechanisms, as is explained below (and see Figures 19.1 to 19.4) but differences are not solely related to phylogeny. Different environments demand major differences in strategies for reproduction and development, as is shown by comparison of marine, freshwater and terrestrial animals (see Box 7.1 and Chapter 13). One strategic difference is the quantity of yolk that the eggs contain. Large quantities of yolk feed the embryo for longer, allowing longer protection within the egg with later hatching, as is often necessary in fresh water and on land, but yolk is expensive to make and is heavy and inert, hampering cell division and cell movement. If there is very little yolk, many more eggs can be produced, the embryo can divide faster and gastrulate unimpeded, but then it needs food. Most invertebrate eggs with little yolk hatch very early into a larval form that can find its own food; a few terrestrial invertebrates are viviparous, the young being retained and fed by the mother within her body: this occurs in some insects (for example tsetse flies) and in some species of *Peripatus*, where the young are fed on internally secreted 'milk'. Many invertebrates in various habitats retain and protect developing eggs ('ovoviviparity'), often in a brood pouch. Within a species also, environmental factors can modify development: for example, the polymorphism between females within a species of social insects may be environmentally determined, and temperature changes may determine the rate and even sometimes the kind of development, as in those vertebrates where sex is determined by ambient temperature.

19.3   What is the pattern of cleavage in invertebrates?

Cleavage is the division of the fertilised egg to form a mass of cells, called 'blastomeres'. All the products of cell division in the embryo normally contain identical sets of genes; only some of the genes will be expressed, but the genes themselves do not change during development (see Figure 19.5 for an experiment elegantly demonstrating this in a tadpole). The cytoplasm or the environment therefore has to be the initial source of information leading to cell diversity. As the environment is seldom reliably constant, usually different substances are localised in different regions of the egg cytoplasm. The plane of successive cleavages is constant for a given group of animals: information contained in the cytoplasm is therefore predictably partitioned between the daughter cells by the process of cleavage. The localised substances, long-lasting messenger RNA and some protein, are supplied entirely by the mother. Cleavage therefore occurs under maternal instruction: for some time it is the mother's genes that control all processes of development.

19.3.1 Spiral cleavage

The pattern of early cleavage may reveal phylogenetic affinity. In animals with three cell layers it largely reflects a major division (see Box 5.2). Spiral cleavage is characteristic of protostomes. After the four-cell stage each daughter cell is set vertically above the furrow between the two cells beneath it, forming a spiral as cell numbers increase (Figure 19.1). In *Lymnaea* whether the cleavage spiral coils to left or to right is determined at the stage when only maternal genes are acting. Later the shell coils in the same direction as the cleavage spiral (this explains a long-standing puzzle, as to why the direction of coiling of the shell depends solely on the mother's genes). Many spirally cleaving animals hatch early as a trochophore larva with a characteristic pattern of ciliation, often but not always including a circumferential ring of cilia, the 'prototroch' (Figures 9.5 and 10.3a).

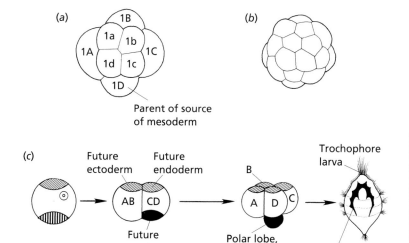

Fig. 19.1 Molluscan early development: (a) spiral cleavage, e.g. *Lymnaea*, 8-cell stage, showing the conventional cell numbering; (b) 16-cell stage of spiral cleavage; both stages viewed dorsally; (c) mesoderm localisation in *Dentalium*.

The first steps of commitment occur at some stage in cleavage. Spirally cleaving eggs were formerly called 'mosaic' eggs because, in an extreme example (ascidians, which are tunicates) separating the first two blastomeres may result in each developing half an embryo, which is striking evidence for early commitment. In these embryos the mesoderm is set aside very early: it can be traced back to a particular cell in the region which will give rise to the endoderm and to nearly all the mesoderm. This blastomere is called '4d', a division product of 'D' in a formal system of labelling the spiral (Figure 19.1a). If the 4d cell is removed a deformed trochophore may develop, lacking mesoderm. Most spirally cleaving animals also have some mesoderm (in the larva and in some adult ectodermal structures) which is ectodermal in origin: it arises from cells 3a and 3b, or occasionally 2b. Acoelomorphs, with a different form of spiral cleavage (see Chapter 6) have endodermal mesoderm arising from 4d but no ectomesoderm.

19.3.2 Radial cleavage

Radial cleavage, where daughter cells are set immediately above the cells below (Figure 19.2), occurs in the deuterostome phyla (echinoderms, hemichordates and chordates) and also in phorona, bryozoans, brachiopods, chaetognaths, rotifers, priapulids and a few others. Experimental work on sea urchins (*Echinus*) in the early twentieth century showed that separation of the first two blasto- meres resulted in regulation of each to form a small complete embryo: this result led to slightly misleading descriptions of radially cleaving eggs as 'regulative', in contrast to the 'mosaic' ascidians and nematodes. Certainly, full commitment does not occur until later in cleavage. Future mesoderm may be defined earlier than used to be thought, but is not apparent at the cleavage stage; it arises later by out-pouching from the gut (Figure 19.2). Larvae of radially cleaving animals (for example echinoderm larvae, Figure 17.8) are distinctively different from a trochophore.

There is evidence that radial cleavage is primitive among animals (see Chapter 20). Spiral cleavage probably evolved once only, so that minor differences in the mechanics in different groups may be significant indicators of phylogeny.

19.3.3 Other cleavage patterns

Among both sponges and cnidarians there are many different cleavage patterns. The ctenophore plate and the acoelan bispiral (mentioned in Chapters 5 and 6) are unusual arrangements of the first blastomeres, and the 'nematode T' is shown in Figure 19.3. Cleavage is asymmetrical in a number of animals, but most strikingly so in nematodes: the blastomeres differ in size, containing different amounts of cytoplasm. This correlates with determination of cell fate by individual cells rather than groups of cells.

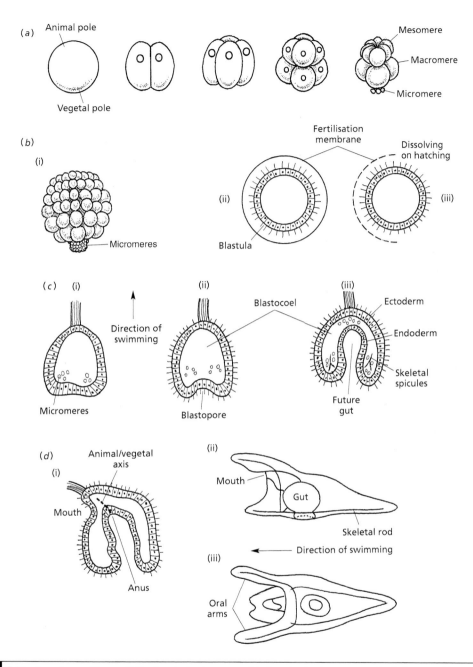

Fig. 19.2 Echinoderm (sea urchin, e.g. *Echinus*) early development, with the animal pole uppermost in all diagrams: (a) successive stages of radial cleavage; (b) blastula, 64-cell stage (i) lateral view, (ii) section of blastula within fertilisation membrane, (iii) hatching; (c) gastrulation, (i) about to start, (ii) blastopore appears, (iii) future gut appears; (d) larva formation, (i) diagram of early larva, (ii) pluteus larva in side view, (iii) pluteus larva in ventral view.

In all the patterns of cleavage described above, the end product (blastula) is a ball of cells, hollow because the cells are secreting fluid into the centre. However animals with large quantities of yolk impeding cleavage, such as most arthropods and many other

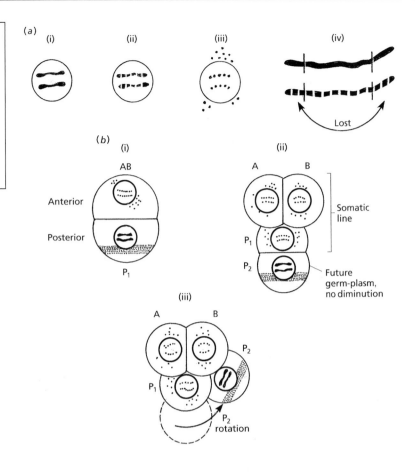

Fig. 19.3 Nematode (*Ascaris*) early development: (a) chromosome diminution, (i) two large chromosomes, (ii) to (iv) fragmentation and loss of one chromosome; (b) the first stages of cleavage, (i) separation of AB from P$_I$, (ii) the four-cell stage with the 'nematode T', (iii) the final form of the four-cell stage.

invertebrates, have very different structures at the end of the process. The yolk is left undivided outside the developing embryo, a condition familiar from hens' eggs where the blastula is a flattened disc lying on top of a mass of yolk. Most arthropod eggs have a large quantity of central yolk, with the embryonic cells peripheral to it. In insects the fertilised egg nucleus divides rapidly many times to form a syncitium, each daughter nucleus having a small surrounding mass of cytoplasm with no cell boundary. The nuclei with adhering cytoplasm migrate from the centre to the periphery. where cell boundaries form. Figure 19.4 shows these stages in *Drosophila*.

19.4 How do invertebrates gastrulate?

The cell movements constituting gastrulation vary considerably among sponges and cnidarians, but the process is very similar in most animals with three cell layers. Most commonly, cells move from the surface of the blastula to the interior by 'invagination', the infolding of a sheet of cells (Figure 19.2c). At a point on the blastula surface called the blastopore some cells move inward, largely

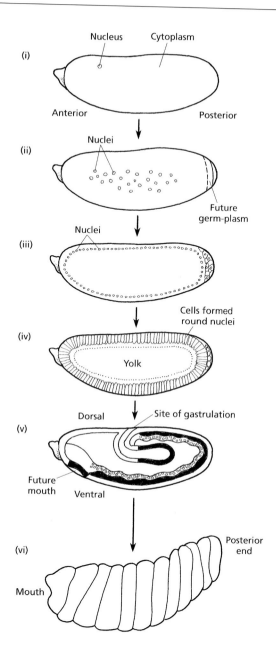

(i)

Nucleus   Cytoplasm

Anterior

Posterior

(ii)

Nuclei

Future
germ-plasm

(iii)

Nuclei

Cells formed
round nuclei

(iv)

Yolk

Dorsal   Site of gastrulation

(v)

Future
mouth   Ventral

Posterior
end

(vi)

Mouth

Fig. 19.4   The early development of *Drosophila*: (i) the egg; (ii) syncitium formed; (iii) nuclei at the periphery; (iv) cytoplasm rounds up around nuclei, making cells; (v) gastrulation; (vi) external view of segmented larva.

obliterating the lumen of the blastula and making an internal layer, the endoderm, which encloses the future gut cavity. The blastopore forms the mouth, and usually by a connecting groove also the anus, in protostomes such as molluscs and nematodes. In deuterostomes such as echinoderms, the blastopore becomes the anus only and the mouth opens separately later. Cleavage of very yolky eggs produces a plate of cells, not a ball, but the invagination process may be very similar; for example, in insects gastrulation occurs within a thickened band of the superficial cells (Figure 19.4). An alternative

the current focus is on identification of genetic mechanisms underlying the evolution of animal design.

While it is exciting to discover how development is initiated and directed by gene action, the process cannot entirely be explained in these terms. In biology reduction to the simplest components can never give us a full picture. The behaviour of cells, like the behaviour of whole organisms, cannot entirely be predicted from the behaviour of the genes that they contain. Interactions between the component parts, and between the developing animal and its environment, will always affect the end product.

Chapter 20

Invertebrate evolutionary history

When invertebrate animals were introduced in earlier chapters of this book they were placed in phyla, according to their body plans, with an indication of the diversity occurring within each phylum. To this extent they were classified, that is, they were ordered into a series of groups. Classification is the basis for phylogeny, which is the tracing of the evolutionary history that links these groups: classification and tracing phylogeny are two distinctively different processes. This concluding chapter uses evidence from genes to trace phylogeny and indicates our present understanding of the course of evolution.

20.1 How can we trace the course of evolution?

As was explained in Chapter 2, evidence from morphology, fossils and molecules (mainly genes) can be combined to trace evolutionary history. The primary difficulty with morphology is that most of the characters in animals have evolved more than once, so that resemblance due to shared ancestry (homology) is difficult to distinguish from resemblance due to similar adaptations to a way of life (convergence), and only homology can help us to trace phylogeny. Since the late 1980s molecular biology has provided new tools for assessing the pattern of evolution. Molecular evidence is not inherently preferable, but it is valuable because it can be compared with morphological evidence from living animals and from fossils. The molecules concerned are mainly genes; study of present-day organisms reveals genetic change, and resistance to change, that is exceedingly informative.

Since the early 1990s there has been a vast outpouring of papers about molecular phylogeny. The study of development has become the study of its control by genes, and understanding of developmental change has contributed to our understanding of phylogeny.

20.2 How have genes provided enough raw material for evolution?

The simple picture of genes and evolution given in Chapters 1 and 2 must now be elaborated, to account for the amount of diversity that has arisen since the Cambrian explosion. We know that changes in the timing of developmental events can produce sudden change (e.g. paedomorphosis can occur, see Chapter 18), but such events are only a small part of the answer. As so often in biology, further enquiry or new knowledge means that the initial account must be amplified and qualified. Yet the simple picture remains an indispensable foundation for further understanding.

Genes, or at least the set of proteins that they encode, are remarkably conservative – that has been one of the main surprises of the last two decades. During evolution many genes have become duplicated, rearranged or lost, but the genes themselves have not changed greatly. Change in gene regulatory networks emerges as the main cause of evolutionary change: regulatory genes specify 'transcription factors', which are proteins acting on short sequences of DNA called 'promoters', situated near to the genes. Gene action depends on regulation of whether a gene is expressed and, if so, at exactly what time during development and in which parts of the body. Gene regulation produces the raw material of evolution in abundance.

Another surprise was the discovery that only about 2% of our DNA codes for proteins; 98% is what used to be called 'junk DNA' (these figures are for humans, but the phenomenon is general). Non-coding DNA is not of course junk, it provides much of the regulation of gene expression. Remarkably, much of this non-coding DNA consists of 'jumping genes' (transposons) that can leave their chromosomes and paste themselves elsewhere in the genome – perhaps beside a promoter, altering its action; some even have their own promoters and may regulate any genes nearby. In these ways transposons can alter gene expression and change the course of evolution.

Even worse – the simple picture of random mutations has to be modified. It has long been known that bacteria can modify the speed at which mutations appear, according to the composition of the medium in which they are cultured: it now appears that animals also can modify the mutation rate – mutations are more likely in some parts of the genome than in others. Genomes contain information that can focus mutations in certain areas and direct it away from others. This means that natural selection can act on the very mechanisms that copy and repair the DNA itself.

All this does not mean that Darwin was wrong, or that what we believe about natural selection is mistaken: only that our picture of selection acting on entirely random mutations is oversimple.

20.3 How can genes help us to trace evolution?

The finding that genes have changed remarkably little during evolution has made an opening for molecular studies of phylogeny: where change is very rare, the number of maintained changes becomes significant. Since phyla with very different gene content are likely to have diverged over a long period of time, studies of the genomes of present-day animals may reveal evolutionary relatedness. This idea was first used to study the genes specifying ribosomal RNA, which are very strongly conserved (perhaps any change would be selectively disadvantageous); other genes are now used as well, including several genes for 'housekeeping' proteins (including elongation factors and tubulin), Hox genes and mitochondrial genes.

Genes directing development also have changed particularly little during evolution: what has changed is the downstream network of regulatory genes under their control. Studies of the Hox genes that pattern development along an axis (see Chapter 19) can help to reveal the course of evolution. Apparently all Bilateria had an ancestral cluster of 5–10 Hox genes. During evolution new genes have emerged, primarily as a result of gene duplication, or become lost. The main bilaterian groups (deuterostomes, ecdysozoa and lophotrochozoa, see below) are characterised by different Hox genes. Like 18S rDNA and 28S rDNA, Hox genes have changed very slowly over time, but they are a quite separate line of molecular evidence: while studies based on ribosomal genes measure sequence divergence, Hox gene studies identify gene sets and potential shared gene duplication.

A third separate line of work that has been developed more recently uses the mitochondrial genome, either by comparing gene order or by sequencing the whole of these small genomes. Many more genes are now being used to check, and often consolidate, the earlier molecular results.

20.4 What do genes tell us about relationships between the earliest phyla?

Research on Hox and other developmental genes is exceedingly active, so that any publication of the results will become out of date very soon. Nevertheless, here is the picture of the course of animal evolution that is widely accepted early in the twenty-first century.

20.4.1 The origin of animals
Animals arose among the Eukaryotes (organisms with cells having a distinct nucleus enclosed by a double membrane) and are more closely related to fungi than either kingdom is to plants.

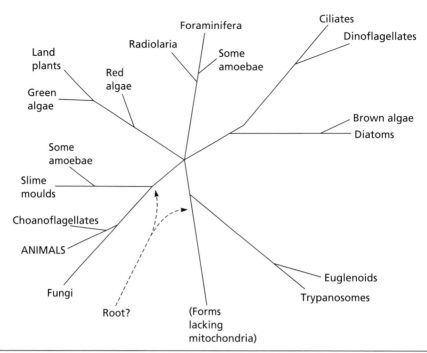

Fig. 20.1 A present hypothesis of the relationships between kingdoms of Eukaryotes.

Figure 20.1 represents one possible sequence; or animals and fungi may have separated much earlier. Our ancestors were unicellular, performing all their functions in a single cell. Such an organism may have a number of characteristics which most animals require, such as motility, sensory receptors, digestion of food, sexual processes. However, to make a multicellular form cells must also be able to stick together, to communicate and form patterns and (almost always) to differentiate to achieve division of labour.

20.4.2   Metazoa are monophyletic

All the multicellular animals we know today had a common ancestor ('Metazoa' becomes synonymous with 'animals'). This is suggested by the uniform molecular constitution of the extracellular matrix, the ubiquity of collagen and many other biochemical features, as well as by 18S rDNA and many shared genes found only in Metazoa. Porifera (sponges) and cnidarians did not evolve quite separately from other animals, as was once believed.

20.4.3   Porifera

Porifera are morphologically the simplest multicellular animals. Since adult sponges have no axis, no Hox genes can be expected, yet homeobox-containing genes are found even here, as indeed they are in yeast; their function in sponges is not clear. Morphological evidence further suggests that the colonial unicellular choanoflagellates (see Chapter 3 and Figure 3.4) may be the forerunners of sponges; this too is supported by molecular evidence: choanoflagellates produce a surprising number of proteins not previously isolated from non-Metazoa, including key cell-signalling and cell-adhesion proteins, and also express genes not previously found in non-metazoans. Later in evolution these genes become co-opted for development.

20.4.4   Cnidaria

The Anthozoa are now confirmed as the basal class in the phylum (see Chapter 4), and molecular evidence suggests that they have arisen from Porifera. A surprising number of genes known from bilateria are active in Cnidaria, often with rather different functions. Compared to other phyla, cnidarians may reveal intermediate stages in the evolution of Hox gene clusters; compared to *Drosophila*, cnidarians have only the anterior and posterior classes of Hox genes. In the highly polymorphic colonies of *Hydractinia* (see Chapter 4) a Hox-like gene is expressed only along stolon and polyp axes and appears to regulate the contents of the colony; studies of regenerating *Hydra* reveal genes with expression patterns parallel to those of genes known to regulate head formation in bilaterians. This finding justifies calling the oral end of *Hydra* a 'head'. In general, genetic evidence shows that cnidarians are more like other animals than used to be thought, and possess remarkably many genes shared with mammals; many of these genes are lost in early Bilateria.

20.4.5   Ctenophora

This phylum (see Chapter 5) remains particularly hard to place: not close to Cnidaria despite superficial appearances, possessing mesoderm yet not like other triploblasts, they remain enigmatic, a separate early offshoot in animal evolution. Genetic evidence places them as probably the most primitive animals after sponges.

20.5   How do genes relate the protostome phyla?

Protostomes are by far the larger of the two traditional groups of the Bilateria (triploblastic animals). There are anomalies and apparent intermediates in the early developmental features that originally defined the division (see Box 5.2), but it is upheld by genetic evidence, which is used to make a further division of the protostomes. Deuterostomes include only very few phyla.

20.5.1   Acoelomorpha

Acoela and Nemertodermata, two groups which share several unusual characters, are uniquely simple among Bilateria (see Chapter 6). But is this simplicity primitive or secondary? Morphological study cannot answer this question. The unique early development of acoelans, said to be shared by nemertodermatids, suggests that they are primitive. It would be hard to derive the acoelan bispiral, with no quartet formation, from any other spirally cleaving protostome, but genetic evidence is needed.

This evidence comes from several genes: use of 18S rDNA is hampered by 'long branches' (rapid nucleotide substitution) but 28S rDNA gives clearer results; both these genes measure the degree of sequence divergence over time. Supporting evidence identifies sets of genes common to both phyla, for example Hox genes, interestingly intermediate in complexity between those of cnidarians and bilaterians, a very different gene coding for nuclear proteins and another coding for certain sequences in myosin. This last is a particularly useful indicator of primitive groups, as it appears to change very regularly over evolutionary time. While Cnidaria have only anterior and posterior clusters of Hox genes, Acoela have three (anterior, posterior and central) as opposed to 8–10 clusters in other bilaterians.

These genes all support not only the combination of acoelans and nemertodermatids as acoelomorphs but also, more importantly, the primitive position of these animals at the base of the Bilateria. Further confirmation of the primitive nature of acoelomorphs is arriving fast, using a number of different genes. That these animals are primitive is important, as a possible evolutionary origin for bilaterians is suggested. In the absence of any simple primitive form, some earlier theories had disregarded the prevalence of convergence and endowed the earliest bilaterians with all the elaborations common to protostomes and deuterostomes. Now the favoured theory is that something like a cnidarian planula gave rise to the earliest acoelomorphs and to the ancestors of all bilaterians, but there are still many alternative views.

20.5.2 Platyhelminthes

All sources of genetic evidence suggest that the Turbellaria do have a common ancestor (but they are not monophyletic because the parasitic platyhelminthes arose from within the group). Genetic evidence supports morphology in placing the Catenulida within the phylum but at its base. They are the sister group of all other Platyhelminthes (Rhabditophora and the parasitic groups). With the acoelomorphs removed, the monophyletic identity of the phylum is restored.

The traditional view of platyhelminths as basic triploblastic animals arose originally because their simplicity was assumed to be primitive. As was explained in Chapter 6, such a view overlooks the specialised morphology of these animals. Like acoelomorphs, platyhelminths have been hard to place because many show very rapid evolution of ribosomal genes ('long branches' on the constructed trees). Evidence from 18S rRNA genes rests on subjective selection of species with slow evolution ('short branches'). Such evidence did place the platyhelminths among the protostome coelomates rather than at the base of the Bilateria, but needed confirmation.

Evidence from a number of genes has confirmed this placing: for example, the occurrence in platyhelminths of Hox genes Lox4 and Lox5, known from polychaetes and leeches and phyla related to annelids.

That platyhelminths are close to coelomates of course tells us nothing about the direction of evolution: it certainly does not imply that platyhelminths are coelomates that have lost their body cavity. There is no evidence that the ancestors of platyhelminthes ever had a coelom: there is now a great deal of evidence to the contrary. Platyhelminths are a separate phylum, outside the annelids but included in the same subdivision of the protostomes.

20.5.3 Nemertinea

Particularly thorough genetic investigation of nemertines was prompted by theories, now abandoned, that they might be deuterostomes on the line to chordates. Evidence from three ribosomal genes, Hox genes, mitochondrial gene order and two other genes combines to place nemertines close to the annelid group of coelomate phyla, and reassessment of their development (see Chapter 7) gives this placing further support. Again, this finding does not mean that an ancestral coelom gave rise to the nemertine rhynchocoel (a restricted cavity which does not even develop in the manner of a protostome coelom); nor that reduction of a coelom produced the unique nemertine blood system (even though it is made from mesoderm which splits to make the lumen of the blood vessels). Molecular evidence can establish relationships but cannot tell us anything about the direction of evolution. Support for the claim that nemertines are reduced coelomates (for example by way of paedogenesis of a small ancestral larva) would require much evidence that at present does not exist and, as was explained in Chapter 7, coelom loss is very unlikely in a phylum consisting mainly of active predatory animals.

20.5.4 Annelida

Annelids are quintessential protosomes. The former combination of annelids with arthropods as 'Articulata', because both are segmented, had been abandoned by most morphologists and is not supported at all by genetic evidence (from ribosomal genes, Hox genes, and mitochondrial gene order). Segment identities are correlated with domains of Hox gene expression. These domains are large and overlapping in most polychaetes, with corresponding genes expressed more anteriorly in leeches. Where there is a very diverse segmental pattern in a polychaete, as in the tube-living *Chaetopterus*, Hox genes are expressed earlier in the larva, followed by posterior discontinuities related to morphological boundaries.

Genetic evidence suggests that annelids are not monophyletic: polychaetes are even more diverse than had been realised. Of the coelomate 'phyla' in Figure 5.4, only Sipuncula with certainty still ranks as a phylum. Pogonophora (Siboglinidae) and Echiura (and possibly Sipuncula) are not separate phyla but arose from within the polychaetes. Traces of segmental ganglia in the larval central nervous system of Echiura support the idea that segmentation has been lost in this group during evolution.

20.5.5 Mollusca

The traditional placing of molluscs in the same group as annelids, on account of similarities in early embryology and larval form, is supported both by newly found fossils and by genetic evidence. The remarkable differences between cephalopods and the other molluscs (from which they evolved) are accompanied by recruitment ('co-option') of existing Hox genes for new developmental functions, as was shown for example in a squid by studies of developmental expression patterns for eight Hox genes.

20.5.6 Phoronida, Bryozoa and Brachiopoda

Radial cleavage and regulative development suggest deuterostome affinities for these animals, yet subsequent features of their early development are either ambiguous or clearly protostome (see Chapter 16). The suggestion that radial cleavage may be primitive among metazoans, with spiral cleavage a secondary specialisation, further devalues a diagnosis based on radial cleavage alone.

Combination of fossil and molecular evidence can resolve this uncertainty about phylogeny. Brachiopods are very common in the fossil record, and resemblances to various early protostomes abound: for example, unlike deuterostomes, brachiopods have chitin, and their mantle-edge bristles are identical in ultrastructure to polychaete chaetae. There are Lower Cambrian fossils called halkieriids which are in several ways intermediate between annelids and molluscs on the one hand and brachiopods on the other. One intriguing slug-like halkieriid from north Greenland has scales like the mollusc *Aplysia*, with a larger shell at either end of the animal; it is tempting to picture this form contracting to make a brachiopod, but such fantasies require evidence. While convergence cannot be ruled out, fossils suggest that annelids, molluscs and brachiopods are closely related. Molecular investigation of modern animals, using the ribosomal genes, mitochondria gene arrangement and Hox genes, strongly supports protostome placing for brachiopods, bryozoans and phoronids. For example, a posterior Hox gene group from the brachiopod *Lingula* is unlike any deuterostome genes and not close to arthropod ones, but is very similar to posterior Hox genes in the annelid ragworm *Nereis* and the limpet *Patella*. All three phyla of animals with lophophores are now placed close to the annelids and molluscs.

20.5.7 Arthropoda

Arthropod monophyly and origins have been particularly controversial. Morphological study of living groups can provide arguments either for a single origin of arthropods (monophyly) or separate origins for crustaceans, chelicerates, insects and others (polyphyly).

Fossil evidence strongly supports monophyly. For example, the biramous limbs of crustaceans were supposed to set them apart from
arthropods with unbranched limbs, which were united as 'Uniramia'
(including onychophorans, myriapods and insects). However, study
of Permian fossils now reveals that the limbs of the earliest arthropods had many branches, the outer ones often subsequently lost.
Modified outer branches of the primitive limb can be traced in
most groups of arthropods, supporting the body wall (in trilobites,
chelicerates and others), forming gills (in aquatic chelicerates and
crustaceans) and, most strikingly, forming the wings of insects.

We do not know which arthropods are the most primitive, but
even though the grouping 'Uniramia' has been disproved, the
closest non-arthropods appear to be the Onychophora (represented
today by terrestrial *Peripatus*, Figure 12.8). Recent fossil finds in
Greenland and south China suggest that the stem groups of both
arthropods and onychophorans may be traced back to Cambrian
marine animals called 'lobopods', whose general body form is not
unlike that of *Peripatus* except for the presence of gills.

Molecular evidence from many sources strongly supports a single
origin for arthropods and also supports the origin of arthropods
with Onychophora from aquatic lobopods.

20.5.8 Nematoda

Most surprisingly, nematodes are now placed near the arthropods,
reviving an ancient grouping of animals with hard cuticles that
need to moult, and use ecdysone in the process. This grouping
does not rely only on a selection of nematodes with slowly evolving
ribosomal genes: it is strongly supported by the study of several
Hox genes, which are known in great detail both for the nematode
Caenorhabditis elegans and for the arthropod *Drosophila melanogaster*.
For example, *C. elegans* shares genes (*Ubx* and *Abd-B*) otherwise
known only from arthropods and Onychophora; further, no arthropod or nematode has been found to share the annelid *Lox* genes,
nor the single *Ubd-A*, found in platyhelminths, annelids, nemertines,
molluscs and protostomes with lophophores.

Following all the above evidence, protostomes are divided into
two groups:

Lophotrophozoa include annelids, molluscs, platyhelminths,
nemertines and protostomes with lophophores (phoronids,
bryozoans and brachiopods).
Ecdysozoa include arthropods, onychophora, tardigrades and
nematodes.

20.6   Where do the smaller protostome phyla fit in?

The smaller phyla without body cavities (or at least without 'true
coeloms') must now be re-examined: the union of many small phyla

as 'pseudocoelomates' has long been considered meaningless, as was explained in Chapter 5. Molecular evidence provides some hope of positive information about their relationships. The phyla introduced in Chapter 5 are at present placed as follows, mainly from 18S rDNA evidence.

20.6.1   Lophotrochozoa

This group includes **Rotifera**, **Acanthocephala**, **Gastrotricha** and **Gnathostomulida** (some or all of these phyla are perhaps the nearest to the platyhelminths). An interesting additional phylum in this group is the misnamed **Mesozoa**. The dicyemids (rhombozoans, Figure 5.2a) can be related to Lophotrochozoa both by ribosomal genes and by one Hox gene. Clearly these very simple endoparasites are not basal metazoans but secondarily specialised. **Myxozoa**, another group formerly misdiagnosed as protistans, are similarly revealed as greatly simplified triploblasts, containing Hox genes (though the function of these genes in animals with no body axis is obscure).

The enigmatic endoparasite *Buddenbrockia* has also been assigned to this group, on evidence from the same families of genes. Molecular evidence is clearly our only tool for placing small simple endoparasites with 'worm-like' features.

Chaetognatha form a small phylum notoriously hard to place: their morphology is unique, use of ribosomal genes is frustrated by very rapid nucleotide substitution and the mitochondrial gene order varies between species. In 2004, however, complete sequencing of the (small) mitochondrial DNA of two chaetognath species has been reported, with results that place the phylum unambiguously among the protostomes, not particularly near to the protostome/deuterostome divide. Further, the mitochondrial genome shares a distinctive amino acid sequence with the deuterostomes, supporting the idea that deuterostome development is primitive and the spiral cleavage and other features of protostome development are derived. The same study sheds further light on the relationship between these two groups, because chaetognaths and other protostomes have a particular amino acid sequence that is found neither in deuterostomes nor in more primitive animals such as anemones. This suggests that deuterostome development may be the more primitive, and protostome development the more derived, condition.

20.6.2   Ecdysozoa

This group appears to include all the moulting 'aschelminthes', that is **Nematomorpha** as well as nematodes, **Kinorhyncha**, and probably **Loricifera**. **Priapula** are identified as belonging to this group not only by various nuclear genes but also by exciting new studies of the mitochondrial genome, where the gene order closely resembles that of arthropods (M. J. Telford, personal communication).

The resulting tree is summarised in Figure 20.2

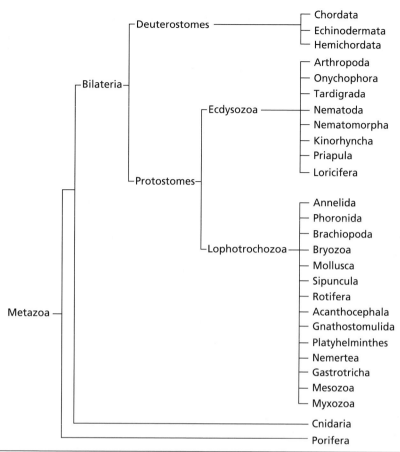

Fig. 20.2 A recent scheme of invertebrate evolution (Ctenophora, Acoelomorpha and Chaetognatha omitted).

20.7   How do genes relate the deuterostome phyla?

The deuterostome phyla are **Echinodermata**, **Chordata** and **Hemichordata**. Morphological evidence can be found to support any pairing of these three phyla, but molecular evidence overwhelmingly supports a close relationship between Echinodermata and the unfortunately named Hemichordata.

20.7.1   Echinoderms

Echinoderms are very unlike most other animals. They can therefore demonstrate some general properties of genes in evolution. In particular, Hox genes familiar from *Drosophila* have been found, expressed in ways specific to echinoderms. Why animals lacking a main body axis should require Hox genes is far from obvious: the order of expression follows a circular route round the baby adult from arm's tip to arm's tip, but how and why this arrangement arose is open to conjecture. Three of these Hox genes appear to be responsible for several of the unique features of the phylum. Another general property of genes demonstrated by echinoderms is 'recruitment', the use of old genes for new roles. Genes regulating development have been shown to be extraordinarily flexible; the expression of echinoderm genes downstream may be unique to the phylum and different in the different classes, and the genes may even be expressed at different regulatory levels. For example, the 'downstream' gene *distal-less* (*Dll*) is well known to be necessary for the distal part of the limb to develop in *Drosophila*: among echinoderms, *Dll* is expressed in the tube feet of asteroids, echinoids and holothuroids but not in those of ophiuroids, where a different gene (also known from *Drosophila*) controls development of tube feet. In asteroid larvae, *Dll* is expressed in ectodermal cells and in particular in the brachiolar arms, used for attachment. In echinoids *Dll* is expressed from the first appearance of the adult rudiment and specifies most of the structures on the oral side, spines as well as tube feet. This is a clear example of recruitment of genes to serve in quite new ways, differing in different echinoderms. It is also an example of unchanging Hox genes acting on different targets owing to 'downstream' changes. A more practical consequence of this flexibility is that these downstream genes may be quite unreliable for indicating relationships.

20.7.2 Hemichordates

Morphological evidence suggests that hemichordates are more closely related to echinoderms than to chordates, an idea early supported by evidence from 18S rDNA genes and more recently confirmed by DNA data from the whole mitochondrial genome of *Balanoglossus*. At once the question foreshadowed in Chapter 18 becomes central: which of the apparent 'chordate characters' are really deuterostome characters that have been lost in echinoderms? This may apply to gill slits, as is suggested by the finding of them in several fossil echinoderms and supported by evidence from certain *pax* genes, which are expressed similarly in the pharynges of hemichordates, urochordates and cephalochordates. This uniformity of development suggests that hemichordate gill slits are not a product of convergence but are the continuation of a primitive deuterostome character.

Morphology suggests homology between the hemichordate and echinoderm nervous systems, with little relation to the chordate hollow double nerve cord. The tornaria larva has a diffuse nerve net very similar to that of echinoderm larvae which (unlike the echinoderm nerve net) continues into the adult with little change. Yet in the adult enteropneust *Saccoglossus kowalevski* this diffuse nerve net expresses the same genes, in the same order, as in the highly developed chordate central nervous system. This extends the morphological evidence by suggesting that the deuterostome ancestor had a diffuse nervous system, later centralised in the chordate lineage.

The morphology of the notochord, however, is known only in chordates. Morphological study showing that the stomocord of hemi-chordates is quite different from a notochord is confirmed by the gene *brachyury*, which is expressed in the early embryonic ectoderm at the site of the developing mouth, in hemichordates and in nearly all bilaterians (and even in the planula of *Hydra*). In chordates, by contrast, *brachyury* is expressed in the posterior mesoderm and is closely associated with notochord and tail formation, confirming the morphological finding that the notochord is unique to chordates.

20.7.3   Chordates

The theory of group origins by neoteny is attractive but hard to prove, and molecular evidence does not so far seem to support it. For example, preliminary 18S rDNA results suggest that larvaceans diverged early from the main stock, and that adult tunicates represent a secondary specialisation rather than an early adult form that has been discarded.

The full genome of the ascidian *Ciona intestinalis* has now been sequenced, in the hope of illuminating both the origins of chordates within the deuterostomes and the origins of vertebrates. The general picture is of genes familiar from other invertebrates (60% are shared with protostomes) but present in much smaller numbers. For example, fibroblast growth factors are governed by one gene in *Drosophila*, two genes in *C. elegans*, six genes in *Ciona* and 22 in mammals. Some genes are unique to the ascidian lineage, such as the genes involved in making cellulose; these resemble metabolic genes in bacteria and fungi and therefore are believed to have been introduced by symbiotic bacteria. Most of the vertebrate gene fami-lies are represented in simplified forms. Homologies with gill slits and gland (thyroid, pineal) are all upheld, but with very different functions for the structures concerned.

Most recently, the larvacean *Oikopleura dioica* has been sequenced, and supports the revolutionary suggestion (based on 146 genes) that tunicates, not cephalochordates, are the closest living relatives of vertebrates. Amphioxus may be closer to echinoderms than to the tunicate/vertebrate line; we now await the sequencing of amphioxus.

The origin and evolution of vertebrates has been illuminated by other genes. The old (1822) suggestion that vertebrates were 'worms turned upside down' with the nerve cord becoming dorsal and the main blood vessel ventral has been rescued from ridicule by study of the genes specifying the dorsoventral axis. The vertebrate increase in complexity has been associated with duplications of Hox gene clusters (for example, amphioxus has one cluster and vertebrates have four). Vertebrates are outside the scope of this book, but study of invertebrate chordates can trace their origins.

20.7.4   *Xenoturbella*

Recently, another deuterostome has been identified. *Xenoturbella* is a small marine worm covered in cilia and very simple indeed in structure. It has no through gut, no body cavity, no brain and no organised gonads. A long-standing enigma, it has been placed as an aberrant flatworm and then as a mollusc, but the molluscan DNA turned out to come from its food. Now the mitochondrial DNA has revealed *Xenoturbella* as a basal deuterostome, its simplicity probably being secondary.

20.8　What do molecules tell us about relationships within phyla?

Chapters 3 to 17 each include a discussion of the relationships between different classes within each phylum, where this can be ascertained, with only brief reference to molecular as well as morphological evidence. Two phyla, echinoderms and arthropods, are here chosen for fuller consideration.

20.8.1　Echinoderms

This phylum is chosen because the fossil record is particularly good for many classes, allowing the comparison of living forms to be illuminated by combining evidence from both fossils and molecules. The unique skeleton can be found in fossils and shows characters known from present-day forms. The earliest known echinoderm fossils, recently discovered in early Cambrian deposits in China, are dated at 680 million years ago (Ma), which is thought to be only 50 Ma after deuterostomes diverged from protostomes. These fossils go back to the roots of echinoderm morphology, before all the distinctive features were clear; these earliest fossils are very interesting but their morphology is difficult to interpret. Fossil evidence from 590 Ma is clearer and suggests possible relationships between the five main classes. Figure 20.3 traces their divergence

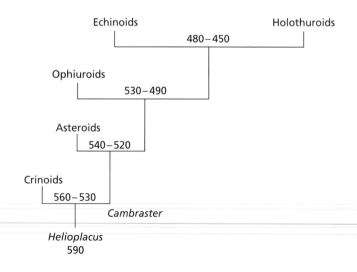

Fig. 20.3 The relationships between echinoderm classes, based on fossil evidence. The numbers represent millions of years ago (Ma) and record the earliest possible times of divergence (higher figures, on left) and the times when divergence has certainly occurred (lower figures, on right).

up to their complete separation, 450 Ma (in the Ordovician). The history of sessile echinoderms, including the crinoids, is well known. Relationships within free-living classes are known best for the echinoids, where fossils are particularly abundant (10 000 fossil species have been described) and where the same characters can be used for classification as in the 900 known present-day species. The rates of morphological change can be estimated, and correlate well with time elapsed, over 250 million years. Ophiuroid trees are well rooted and supported by fossil evidence; asteroid trees much less so, and some authorities place asteroids closer to ophiuroids than Figure 20.3 suggests. Holothuroids, with poorly developed skeletons, have left few fossils and their relationships remain controversial.

Molecular evidence, largely derived from genes specifying ribosomal RNA, has not always given results agreeing with the fossil evidence. More recently, however, Hox genes have been intensively studied and these results can complement and extend the fossil evidence. Relationships within the echinoids have now been clarified by a particularly thorough comparison of information from fossils with that from genes, comparing 163 characters of fossil morphology with analysis of two different ribosomal genes, 15S and 28S rDNA.

20.8.2 Arthropods

Arthropod classes have been chosen for fuller discussion because their great diversity arises from a precisely defined common segmental structure, constrained by an exoskeleton. All arthropods have similar sets of Hox genes, variously expressed. Morphology changes at boundaries between segments or tagmata (head, thorax, abdomen) are associated with discontinuities in Hox gene expression, and the presence or absence of segmental appendages in different arthropods can be traced to Hox gene action. For example, the brine shrimp *Artemia* has six genes of the same Hox family as *Drosophila*, and five of these are very similar indeed in the two animals. Their different pattern of appendages depends upon differences in the expression of these genes.

Insects and crustaceans

Evidence from two different ribosomal genes first suggested that insects and crustaceans have a common root within the arthropods.

This suggestion is supported by evidence from mitochondrial gene order, where the two classes share the same changes from the primitive arrangement. All these genes must have originated before the Insecta and Crustacea diverged (in the Cambrian). Since a common ground plan underlies the segment diversity in insects and crustaceans, it is informative to consider the two groups together.

The head segments of Crustacea correspond to those of Insecta, but the second segment bears a second pair of antennae (a general difference between the two classes): in crustaceans the relevant Hox gene is not expressed in that segment, so that the downstream gene *distal-less* (*Dll*) allows a second pair of antennae to develop.

Many biting and chewing insects and crustaceans have mandibles on the third segment: fossil evidence suggests that the mandible in both classes evolved from the whole limb by truncation of the axis. Neither insects nor crustaceans express *Dll* in this limb, so that only a short stubby appendage can form.

Behind the head, identical Hox genes govern the pattern of appendages in both *Drosophila* and *Artemia*; differences in whether, where or when they are expressed may cause them to have different effects. The main Hox genes acting are *Antennapedia* (*Antp*), *Ultrabithorax* (*Ubx*) and *Abdominal A* and *B* (*Abd-A* and *Abd-B*). *Ubx* and *Abd-B* are ancient genes, shared with Onychophora (the arthropod sister group) and with the nematode *Caenorhabditis elegans*, but not with members of the other group of protostomes. These genes are not concerned with the nature of the limb (e.g. uniramous or biramous) but only with whether and when it is formed. *Dll* is always necessary for limbs to be fully formed (see Chapter 19 for an introduction to all these genes).

In *Drosophila* the thorax has three pairs of limbs and the abdomen has none. *Antp* is expressed in the thorax, promoting limb formation, and *Ubx* acts with *Abd-A* in the abdomen, where *Ubx* represses *Dll* and limb formation is prevented.

In *Artemia* all three Hox genes are expressed together in all segments behind the head and all these segments bear (similar) limbs. This is possible because here *Ubx* can repress *Antp* without repressing *Dll*. Other crustaceans are found to share this more anterior expression of these genes; the flexibility in *Ubx* action is also found to be shared by other crustaceans. In Onychophora, *Ubx* promotes limb formation, suggesting that this is its primitive function. The difference has been traced to a particular peptide which has been lost in the evolution of insects. This accounts for the very constant pattern of insect limbs (three pairs in the thorax, none in the abdomen), in contrast to the great variety found in crustaceans. In both *Drosophila* and *Artemia*, *Abd-B* is expressed in segments forming the gonads.

Crustacean diversity is associated with many different patterns of gene expression: for example, in isopods one of the 'head' genes is expressed in the first thoracic segment, where feeding appendages are formed in place of walking legs; barnacles have lost *Abd-A* as well as the abdomen; in decapods alterations in *Ubx* and *Antp* expression have led to greatly varied limb specialisations.

Ubx also governs wing formation in insects: most insects have two pairs of wings, on the second and third thoracic segments; *Drosophila*, like other Diptera, has only one pair, because *Ubx* transforms the posterior pair into halteres.

Chelicerates

The chelicerate body is uniquely divided into the prosoma and the opisthosoma. Antennae are totally absent, the first segment bears chelicerae, the second bears pedipalps and the last four segments of the prosoma bear the four pairs of walking legs. The unique limb pattern of the prosoma is not due to the absence of any segments but to differences in Hox gene expression: for example no antennae can be formed in the first segment because as a result of such a difference *Dll* is not expressed there, and the siting of chelicerate appendages relates to the confinement of *Ubx* expression to the opisthosoma. In all living chelicerates studied, the expression of Hox genes shows that the prosoma as a whole corresponds to the head of insects and crustaceans. The appendages used as jaws by many insects and crustaceans are used as legs by chelicerates.

Myriapods

In myriapods the most anterior segment bears antennae, while the second has mandibles, where the expression of *Dll* promotes a long jointed mandible in contrast to insects and crustaceans, and there may be maxillae and sometimes a pair of poison claws. All more posterior segments express both *Ubx* and *Abd-A* and have similar walking legs. The old idea that insects were derived from myriapod-like arthropods is not supported by any kind of molecular evidence. Myriapods, like chelicerates, have the same set of Hox genes as insects and all other groups of arthropods; that these genes have not been used to make specialised segments does not relate them to insects in particular, and some other evidence suggests that they may be closer to chelicerates.

Onychophora are found to have exactly the same set of Hox genes as centipedes, with homologues for each gene expressed in *Drosophila*. Whether or not there is specialisation of appendages, the genes are there.

In summary, Hox genes, discovered and first studied in relation to ontogenetic development, can also assist the assessment of homology and hence the investigation of phylogeny. Arthropods are seen to be a monophyletic phylum sharing common Hox genes and having a common origin of segmentation, with early divergence of down-stream genes that later led to the separation of the four main groups. Figure 20.4 indicates a possible phylogeny of these groups.

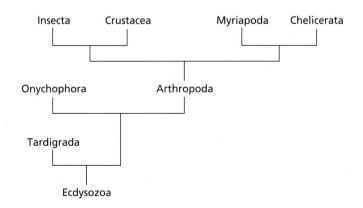

Fig. 20.4 A possible phylogeny for the arthropod classes and related phyla. Alternatively, Insecta may have arisen from among Crustacea.

20.9   Can we now define homology?

Homology, provisionally defined as similarity in morphology reflecting a common evolutionary origin, is said to be the central concept of comparative biology. Can we give it more than a superficial definition? The difficulty indicated at the beginning of Chapter 2 can now be examined.

Historically, homology is a pre-Darwinian concept. The process of classification (ordering) of organisms acquired new significance after Darwin wrote, 'On my theory, unity of type is explained by unity of descent'. The resemblances on which classification is based were seen to be due to descent with modification from a common ancestor. The importance and difficulty of distinguishing between homology and convergence soon became recognised as paramount.

Homology is phylogenetic continuity, recognised where particular structures have been established and maintained by evolution. We must ask, continuity of what? Of structure alone? Developmental process? Gene action?

Structural homology, the sole basis of the original definition, already demands discrimination between different levels of organisation. For example, the wings of birds and bats can be called convergent because they are very differently constructed and, considered at the level of wings, have no common ancestor. Equally reasonably they can be called homologous when considered at the level of the whole animal, because they have evolved from the same pattern of vertebrate forelimb, and that can be traced to a common ancestor. A major problem with structural homology is that if it is used to construct a phylogeny and that phylogeny is then used to deduce homology, the circularity does need to be noticed! A phylogeny needs to be constructed first, on widely based evidence, before homology is postulated.

Development and evolution are now recognised as closely linked: identity of developmental pathways might therefore appear to be an important criterion of homology. However, different developmental processes can produce the same adult form by different routes: examples include methods of gastrulation, and the different developmental pathways of regeneration and embryonic development in the hydroid *Tubularia* (see Chapter 19). There are many such examples, and conversely the same developmental pathways can produce different adult forms. Nor can embryonic developmental origin provide a conserved adult morphology, since that must depend in part on later stabilising interactions between cells. There seems to be no basis in developmental mechanisms for a general guide to homology.

The molecular level is unhelpful because genetic identity does not coincide with morphological identity; gene behaviour is more complicated than that. A striking example of how genes work is given by the *pax-6* gene, which is expressed at the earliest stage of eye development. The protein specified by *pax-6* initiates eye production in animals as different as insects, squids and ourselves, and also in animals with minimally developed eyes such as platyhelminths and nemertines. It specifies not the form of the eye, but the site for some sort of eye to develop. Grafting experiments give a clear picture: for example, insertions of *Drosophila pax-6* (gene or protein) into *Drosophila* wings or limb rudiments can induce eyes in the wrong place. When a squid or even a mouse *pax-6* gene (as cDNA) is grafted into a *Drosophila* wing or appendage, it too can be expressed, but the eye produced is a squid-like or mouse-like eye, not a *Drosophila* eye. Such experiments clearly show that flexibility and change reside not so much in the genes as in the associated networks of regulatory genes that determine whether, when and where genes are expressed.

Pax-6 action provides another example of the need to define the level at which homology is postulated. Arthropod compound eyes have long been considered to have evolved quite independently of the vertebrate simple eye, but this no longer seems likely. The remarkable resemblance between the eyes of cephalopods and vertebrates is quoted as a classic case of convergence: this remains correct, but it is based not on the gene initiating eye development but on genes downstream from it. The gene initiating eye formation is the same in a wide variety of animals, including platyhelminths, molluscs, arthropods and vertebrates, but the eyes resulting from expression of this gene are not at all necessarily homologous.

Disparity between genetic and morphological identity is also caused by co-option of genes during evolution for very different functions: examples of co-option in echinoderm Hox genes have been given above.

A fundamental difficulty in using genes to detect homology is that, as at every level of organisation, homology cannot be equated with similarity but must be linked to a common ancestor. This is very difficult to find at the molecular level. Also there is increasing evidence that convergence has occurred in molecules, but at this level it is particularly difficult to distinguish between primitive and derived states. The genetic changes underlying morphological convergence are not yet understood.

This brief account shows that while recent advances in developmental genetics can build bridges between the genotype and the phenotype, the identification of homology remains a problem. Close investigation seems increasingly to blur the distinction between homology and convergence. Homology was originally a statement of structural pattern, able to be described as resemblance caused by continuity of information so long as our understanding of this information was insufficiently precise.

Comparison of processes as well as patterns must now be involved. Does homology amount to shared expression patterns of the same regulatory genes at a given stage? Can it be reduced to patterns caused by similar proteins related by common ancestry? Such reductionism can only give a very limited picture.

The definition of homology remains very controversial. We probably have to accept that ancient words, especially pre-Darwinian words, may not be useful in the light of present-day knowledge. We must not allow this ancient word to impede important work on the levels of biological organisation in relation to evolution. Superficially, homology can be used to mean similarity in morphology reflecting a common evolutionary origin, but fuller consideration compels the verdict that, in J. Maynard Smith's words, homology is a 'word fit for burning'.

20.10   Conclusion

It is exciting to have new tools for tracing invertebrate relationships. We can now imagine becoming able to trace the course of evolution with some certainty. A record of history, locked in very deep time, is no longer hopelessly inaccessible. However, the study of relationships should not distract us from the animals themselves. Natural selection has given us an enormous variety of animals. Their structure, functioning and behaviour can tell us more about evolution than we can learn from any phylogeny. The greatest challenge now is to conserve these animals in the threatened and threatening world of today. If we cut down rain forests, kill off coral reefs and poison other habitats on land, in fresh water and in the sea, animals will disappear. Humankind, like any other species, cannot live in the absence of many other forms of life. Conservation is the most urgent task for all biologists today.

Further reading

The second edition of this book draws on a very large number of recent publications. In the interests of keeping the book accessible and uncluttered, sources are not generally given within the text.

This reading and reference list is selected for two different purposes. The first objective is to indicate general works of reference, for invertebrates as a whole and for particular topics. As far as possible these recommendations select books and widely available journals such as *Nature*, London, and *Science*, New York; the Internet is also an increasingly valuable resource. There are many large texts on invertebrates, but no book written before the mid 1990s can be up to date on phylogeny.

The second objective of this list is to identify and acknowledge the sources of some of the new discoveries and new ideas about invertebrate relationships, and these references mainly concern Chapter 20. They include elaborate and advanced papers from a wide variety of journals that are not designed for introductory reading.

For the chapters concerning particular phyla, and for Chapters 19 and 20, topics are listed as far as possible in the sequence of their discussion in the chapter concerned.

General reference

E. E. Ruppert, R. S. Fox & R. D. Barnes, *Invertebrate Zoology*, 7th edn (London: Brooks Cole, 2003). A comprehensive and relatively recent textbook, with sections on general principles introducing each phylum.

R. C. Brusca & G. J. Brusca, *Invertebrates*, 2nd edn (Sunderland, MA: Sinauer, 2003).

V. Pearse, J. Pearse, M. Buchsbaum & R. Buchsbaum, *Living Invertebrates* (Palo Alto, CA: Blackwell, 1987). The enlarged successor to Buchsbaum's *Animals Without Backbones*, with illustrations in colour.

C. Tudge, *The Variety of Life: a Survey and a Celebration of All the Creatures That Have Ever Lived*. (Oxford: Oxford University Press, 2000).

R. McNeil Alexander, *Animals* (Cambridge: Cambridge University Press, 1990). The mechanics of animal structure in relation to locomotion.

F. W. Harrison (ed.), *Microscopic Anatomy of Invertebrates* (New York, NY: Wiley-Liss, 1991–1999). Specialist books (20 volumes) on the cellular structure of all invertebrates. Not elementary reading except for useful introductory summaries of some phyla and groups.

P. Willmer, G. Stone & I. Johnston, *Environmental Physiology of Animals*, 2nd edn (Oxford: Blackwell, 2005). A wide-ranging and up-to-date reference book.

General reading

New Scientist, weekly, is warmly recommended.

Scientific American, monthly, has authoritative and beautifully illustrated articles, rather rarely about invertebrates.

S.J. Gould, *Dinosaur in a Haystack* (London: Penguin, 1997), and many other collections of essays.

M.J. Wells, *Lower Animals* (London: Weidenfeld & Nicolson, 1968). Out of print and not up to date, but a pleasure if you can find it.

M.J. Wells, *Civilization and the Limpet* (Cambridge, MA: Perseus Books, 1998). Read about animals, mostly in the sea.

Evolution and genetics (Chapters 1–2)

C. Darwin, *The Origin of Species by Means of Natural Selection: the Preservation of Favoured Races in the Struggle for Life* (London: John Murray, 1859; later editions are available).

R. Dawkins, *The Blind Watchmaker* (London: Longmans, 1986).

R. Dawkins, *The Selfish Gene* (Oxford: Oxford University Press, 1976).

R. Dawkins, *The Ancestors' Tale: a Pilgrimage to the Dawn of Life* (London: Weidenfeld & Nicolson, 2004).

S. Jones, *The Language of the Genes* (London: Flamingo, 1994).

C. Wills, *The Wisdom of the Genes* (Oxford: Oxford University Press, 1991).

C. Tudge, *In Mendel's Footnotes* (London: Jonathan Cape, 2000).

M. Ridley, *The Red Queen* (London: Viking, 1993; reprinted Penguin, 1994).

A.L. Panchen, *Classification, Evolution and the Nature of Biology* (Cambridge: Cambridge University Press, 1992).

R.A. Jenner, Evolution of animal body plans. *Evolution and Development*, **2** (2000), 208–221.

R.A. Jenner, Unleashing the force of cladistics? Metazoan phylogenetics and hypothesis testing. *Integrated and Comparative Biology*, **3** (2003), 207–218.

J. Moore & P.G. Willmer, Convergent evolution in invertebrates. *Biological Reviews*, **72** (1997), 1–60.

Fossils (Chapter 2)

M.J. Benton, Stems, nodes, crown clades, and rank-free lists: is Linnaeus dead? *Biological Reviews*, **75** (2000), 633–648.

M. de L. Brooke, How old are animals? *Trends in Ecology and Evolution*, **14** (1999), 211–212.

G. E. Budd & S. Jensen, A critical reappraisal of the fossil record of the bilaterian phyla. *Biological Reviews*, **75** (2000), 253–295.

R. A. Fortey, *Life, an Unauthorised Biography: a Natural History of the First Four Thousand Million Years on Earth* (London: Flamingo, 1998).

S. Conway Morris, The fossil record and the early evolution of the Metazoa. *Nature*, **361** (1993), 219–225.

S. Conway Morris, Eggs and embryos from the Cambrian. *BioEssays*, **20** (1998), 676–682.

S. Conway Morris, *The Crucible of Creation: the Burgess Shale and the Rise of Animals* (Oxford: Oxford University Press, 1998).

A. H. Knoll, Breathing room for early animals. *Nature*, **382** (1996), 111–112.

Invertebrate phyla (Chapters 3–18)

Porifera (Chapter 3)

J. Vacelet & N. Boury-Esnault, Carnivorous sponges. *Nature*, **373** (1995), 333–335.

S. P. Leys & G. O. Mackie, Electrical recording from a glass sponge. *Nature*, **387** (1997), 29–30.

S. P. Leys & B. M. Degnan, Cytological basis of photoresponsive behavior in a sponge larva. *Biological Bulletin*, **201** (2001), 323–338.

A. Ender & B. Schierwater, Placozoa are not derived cnidarians: evidence from molecular morphology. *Molecular Biology and Evolution*, **20** (2003), 130–134.

O. Voigt, A. G. Collins, J. S. Pearse *et al.*, Placozoa: no longer a phylum of one. *Current Biology*, **14** (2004), R944–R945.

Symposium (2005) (S. Nichols & G. Worheide), Sponges: new views of old animals. *Integrated and Comparative Biology*, **45** (2005), 333– (first few papers).

Cnidaria (Chapter 4)

U. Frank, T. Leitz & W. A. Muller, My favourite animal: the hydroid *Hydractinia versatile*, an informative cnidarian representative. *BioEssays*, **23** (2001), 963–971.

M. M. Coates, Visual ecology and functional morphology of Cubozoa. *Integrated and Comparative Biology*, **43** (2003), 542–548.

K. Nordstrom, R. Wallen, J. Seymour *et al.*, A simple visual system without neurons in jellyfish larvae. *Proceedings of the Royal Society of London B*, **270** (2003), 2349–2354.

D. E. Nillson, L. Gislen, M. M. Coates *et al.*, Advanced optics in a jellyfish eye. *Nature*, **435** (2005), 201–205.

A. C. Baker, C. J. Starger, T. R. McClanahan *et al.*, Corals' adaptive response to climate change. *Nature*, **430** (2004), 741.

J. M. Pandolfi, R. H. Bradbury, E. Sala *et al.*, Global trajectories of the long-term decline of coral reef ecosystems. *Science*, **301** (2003), 955–958.

T. A. Gardner, I. M. Cote, J. A. Gill *et al.*, Long-term region-wide declines in Caribbean corals. *Science*, **301** (2003), 958–960.

C. Wild, M. Huettal, A. Klueter *et al.*, Coral mucus functions as an energy carrier and particle trap in the reef ecosystem. *Nature*, **428** (2004), 66–70.

Symposium (2003) (R.A. Dewel), New perspectives on the origin of metazoan complexity. *Integrated and Comparative Biology*, **43** (2003), 1–86. Note useful papers on hexactinellid sponges, epithelium, the Cambrian fossil record, and also the following:

P. Cartwright, Developmental insights into the origin of complex colonial Hydrozoa. *Integrated and Comparative Biology*, **43** (2003), 82–86.

R.M. Rieger & P. Ladurner, The significance of muscle cells for the origin of mesoderm. *Integrated and Comparative Biology*, **43** (2003), 47–54.

D.K. Jacobs & R.D. Gates, Developmental genes and the reconstruction of metazoan evolution. *Integrated and Comparative Biology*, **43** (2003), 11–18.

U. Technau, S. Rudd, P. Maxwell *et al.*, Maintenance of ancestral complexity and non-metazoan genes in two basal Cnidaria. *Trends in Genetics*, **21** (2005), 633–639.

Worms (Chapter 5)

Symposium (2002) (J. R. Garey), The lesser known protostome taxa. *Integrative and Comparative Biology*, **42** (2002), 611–703. Papers on Kinorhynchs, Nematomorphs, Gastrotrichs, Loricifera, Cycliophora, Rotifers, Acanthocephala; also Nemertea (Chapter 7) and Tardigrades (Chapter 12).

Parasitism (Box 6.1)

B.E. Matthews, *An Introduction to Parasitology* (Cambridge: Cambridge University Press, 1998).

K. Clay, Parasites lost. *Nature*, **421** (2003), 585–586.

M.E. Torchin, K.D. Laferty, A.P. Dobson *et al.*, Introduced species and their missing parasites. *Nature*, **421** (2003), 628–630.

R.M. Hoek, R.E. van Kesteren, A.B. Smit *et al.*, Altered gene expression in the host brain caused by a trematode parasite: neuropeptide genes are preferentially affected during parasitosis. *Proceedings of the National Academy of Sciences*, **94** (1997), 14072–14076.

L.D. Hurst & J. Randerson, Parasitic sex puppeteers. *Scientific American*, April 2002, 42–47.

Nemertea (Chapter 7)

R. Gibson, *British Nemerteans* (Cambridge: Cambridge University Press, 1982).

Sea, fresh water and land (Box 7.1)

C. Little, *The Terrestrial Invasion* (Cambridge: Cambridge University Press, 1990).

Nematoda (Chapter 8)

J. Hodgkin, H.R. Horvitz, B.R. Jasny & J. Kimble, *C. elegans*: sequence to biology. *Science*, **282** (1998), 2011. Introduction to a special issue of *Science* devoted to *C. elegans*.

R.H.A. Plasterk, The Year of the Worm. *BioEssays*, **21** (1999), 105–109.

M. Blaxter, Two worms are better than one. *Nature*, **426** (2003), 395–396.

P. Cohen, Review of work on RNA interference (RNAi). *New Scientist*, 14 September 2002, 28–33.

Annelida (Chapter 9)

T. Bartolomaeus, Structure, function and development of segmental organs in Annelida. *Hydrobiologia*, **402** (1999), 21–37.

Mollusca (Chapter 10)

R. Martin & P. Walther, Effects of discharging nematocysts when an aeolid nudibranch feeds on a hydroid. *Journal of the Marine Biological Association, UK*, **82** (2002), 455–462.

P. G. Greenwood, K. Garry, A. Hunter *et al.*, Adaptable defense: a nudibranch mucus inhibits nematocyst discharge and changes with prey type. *Biological Bulletin*, **206** (2004), 113–120.

Cephalopoda (Chapter 11)

R. T. Hanlon & J. Messenger, *Cephalopod Behaviour* (Oxford: Oxford University Press, 1998).

Arthropoda (Chapter 12)

G. E. Budd, Why are arthropods segmented? *Evolution and Development*, **3** (2001), 332–342.

G. E. Budd, Tardigrades as stem-group arthropods: the evidence from the Cambrian fauna. *Zoologischer Anzeiger*, **240** (2001), 265–279.

R. A. Fortey & R. H. Thomas (eds.), *Arthropod Relationships* (London: Chapman & Hall, 1997).

S. Barclay, J. E. Ash & D. M. Rowell, Environmental factors influencing the presence and abundance of a log-dwelling invertebrate, *Euperipatoides rowelli*. *Journal of Zoology, London*, **250** (2000), 425–436.

Crustacea (Chapter 13)

J.-Y. Chen, J. Vannier & D. Y. Huang, The origin of Crustacea: new evidence from the early Cambrian in China. *Proceedings of the Royal Society of London B*, **268** (2001), 2181–2187.

D. V. Lavrov, W. M. Brown & J. L. Boore, Phylogenetic position of the Pentastomida and (pan) crustacean relationships. *Proceedings of the Royal Society of London B*, **271** (2004), 537–544.

Chelicerata (Chapter 14)

R. B. Barlow, What the brain tells the eye [in the horseshoe crab]. *Scientific American*, April 1990, 66–71.

Insecta (Chapter 15)

R. F. Chapman, *The Insects: Structure and Function*, 4th edn (Cambridge: Cambridge University Press, 1998).

S. H. P. Maddrell, Why are there no insects in the open sea? *Journal of Experimental Biology*, **201** (1998), 2461–2464.

M. McLeod & S. Braddy, Invasion Earth. *New Scientist*, 8 June 2002, 38–41.

C. P. Ellington, C. Van der Berg, A. Willmott *et al.*, Leading-edge vortices in insect flight. *Nature*, **384** (1996), 626–630.

R. Wootton, How flies fly. *Nature*, **400** (1999), 112–113.

Animals with lophophores (Chapter 16)

T. Bartolomaeus, Ultrastructure and formation of the body cavity lining in *Phoronis muelleri. Zoomorphology*, **120** (2001), 135–148.

L. Peck & D. K. A. Barnes, Metabolic flexibility: the key to long-term evolutionary success in Bryozoa? *Proceedings of the Royal Society of London B*, **271** (2004) (Suppl.), S18–S21.

Echinodermata (Chapter 17)

I. C. Wilkie, Mutable collagenous tissues: extracellular matrix as mechano-effector. *Echinoderm Studies*, **5** (1996), 61–102.

J. Aizenburg, A. Tkachenko, S. Weiner *et al.*, Calcitic microlenses as part of the photoreceptor system in brittle stars. *Nature*, **412** (2001), 819–822.

A. A. Eaves & A. R. Palmer, Widespread cloning in echinoderm larvae. *Nature*, **425** (2003), 146.

M. S. Vickery & J. B. McClintock, Regeneration in metazoan larvae. *Nature*, **394** (1998), 140.

Invertebrate Chordata and Hemichordata (Chapter 18)

T. C. Lacalli, Vetulicolians: are they deuterostomes? Chordates? *BioEssays*, **24** (2002), 208–211.

D. G. Shu, S. Conway Morris, H. Chen *et al.*, Primitive deuterostomes from the Chengjiang Lagerstätte (Lower Cambrian, China). *Nature*, **414** (2001), 419–424.

P. N. Dilly, *Cephalodiscus graptolitoides* sp.nov.: a probable extant graptolite. *Journal of Zoology, London*, **229** (1993), 69–78.

S. Rigby, Graptolites come to life. *Nature*, **362** (1993), 209–210.

C. J. Lowe, M. Wu, A. Salic *et al.*, Anteroposterior patterning in hemichordates and the origins of the chordate nervous system. *Cell*, **113** (2003), 853–865.

D. Tautz, Chordate evolution in a new light [comment on paper by Lowe *et al.*] *Cell*, **113** (2003), 812–813.

E. E. Ruppert, Evolutionary origin of the vertebrate nephron. *American Zoologist*, **34** (1994), 542–553.

G. Mayer & T. Bartolomaeus, Ultrastructure of stomocord and heart glomerulus complex in *Rhabdopleura compacta* (Pterobranchiata) and phylogenetic implications. *Zoomorphology*, **122** (2003), 125–133.

J. Q. Henry, K. Tagawa & M. Q. Martindale, Deuterostome evolution: early development in the enteropneust hemichordate *Ptychodera flava. Evolution and Development*, **3** (2001), 375–390.

Y. Nakajima, T. Humphreys & H. Kaneko *et al.*, Development and neural organisation of the tornaria larva of the Hawaiian hemichordate *Ptychodera flava. Zoological Science*, **21** (2004), 69–78.

Deep-sea invertebrates (Box 18.1)

C. L. Van Dover, *The Ecology of Deep-sea Hydrothermal Vents* (Princeton, NJ: Princeton University Press, 2000).

C.T.S. Little & R.C. Vrijenhoek, Are hydrothermic vent animals living fossils? *Trends in Ecology and Evolution*, **18** (2003), 582–588.

A. Byatt, A. Fothergill & M. Holmes, *The Blue Planet: a Natural History of the Oceans* (London: BBC, 2001). (Based on David Attenborough's television series).

Development (Chapter 19)

L. Wolpert, R.S.P. Beddington, J.P. Brockes *et al.*, *Principles of Development*, 2nd edn (London: Current Biology, 2002).

P.A. Lawrence, *The Making of a Fly: the Genetics of Animal Design* (Oxford: Blackwell, 1992).

A. Minelli, *The Development of Animal Form: Ontogeny, Morphology and Evolution* (Cambridge: Cambridge University Press, 2003).

J.D. Lambert & L.M. Nagy, Asymmetrical inheritance of centrosomally localised mRNAs during embryonic cleavages. *Nature*, **420** (2002), 682–686.

P. Lemaire & S. Marcellini, Early animal embryogenesis. *Biologist*, **50** (2003), 137–140.

G.A. Wray, Punctuated evolution of embryos. *Science*, **267** (1995), 1115–1116.

M.Q. Martindale & J.Q. Henry, Intracellular fate mapping in a basal metazoan, the ctenophore *Mnemiopsis leidyi*, reveals the origins of mesoderm and the existence of indeterminate cell lineages. *Developmental Biology*, **214** (1999), 243–257.

J.-Y. Lee & B. Goldstein, Mechanisms of cell positioning during *C. elegans* gastrulation. *Development*, **130** (2003), 307–320.

B. Goldstein & G. Freeman, Axis specification in animal development. *BioEssays*, **19** (1997), 105–116.

L.M. Angerer & R.C. Angerer, Animal–vegetal axis patterning mechanisms in the early sea urchin embryo. *Developmental Biology*, **218** (2000), 1–12.

J.R. Bayascas, E. Castillo, A.M. Muñoz-Mármol *et al.*, Planarian Hox genes: novel patterns of expression during regeneration. *Development*, **124** (1997), 141–148.

R.A. Raff, Evo-devo: the evolution of a new discipline. *Nature Reviews Genetics*, **1** (2000), 74–79.

G.P. Wagner, What is the promise of developmental evolution? *Journal of Experimental Zoology*, **288** (2000), 95–98.

L. Wolpert & E. Szathma'ry, Evolution and the egg. *Nature*, **420** (2002), 745.

I.C. Scott & D.Y.R. Stainier, Twisting the body into shape. *Nature*, **425** (2003), 461–463.

S.B. Carroll, *Endless Forms Most Beautiful: the New Science of Evo Devo and the Making of the Animal Kingdom* (New York, NY: Norton, 2005).

Invertebrate evolutionary history (Chapter 20)

Genes

R.A. Raff, *The Shape of Life: Genes, Development and the Evolution of the Animal Form* (Chicago, IL: University of Chicago Press, 1996).

S.B. Carroll, J.K. Grenier & S.D. Weatherbee, *From DNA to Diversity*, 2nd edn (Oxford: Blackwell, 2001).

J. W. Valentine, *On the Origin of Phyla* (Chicago, IL: University of Chicago Press, 2004).

M. Levine & R. Tjian, Transcription regulation and animal diversity. *Nature*, **424** (2003), 147–151.

S. M. Rosenburg & P. J. Hastings, Worming into genetic instability. *Nature*, **430** (2004), 625–626.

J. P. Curole & T. D. Kocher, Mitogenomics: digging deeper with complete mitochondrial genomes. *Trends in Ecology and Evolution*, **14** (1999), 394–398.

Origins of Metazoa

S. L. Baldauf, The deep roots of eukaryotes. *Science*, **300** (2003), 1703–1706.

W. Martin & T. M. Embley, Early evolution comes full circle. *Nature*, **431** (2004), 134–137.

A. G. B. Simpson & A. J. Roger, Eukaryotic evolution: getting to the root of the problem. *Current Biology*, **12** (2002), R691–R693.

P. O. Wainright, G. Hinkle, M. L. Sogin *et al.*, Monophyletic origins of the Metazoa: an evolutionary link with fungi. *Science*, **260** (1993), 340–342.

N. King & S. B. Carroll, A receptor tyrosine kinase from choanoflagellates: molecular insights into early animal evolution. *Proceedings of the National Academy of Sciences*, **98** (2001), 15032–15037.

N. King, C. T. Hittenger & S. B. Carroll, Evolution of key cell signaling and adhesion protein families predates animal origins. *Science*, **301** (2003), 361–363.

Earliest present-day phyla

H. Bode, D. Matinez, M. A. Shenk *et al.*, Evolution of head development. *Biological Bulletin*, **196** (1999), 408–410.

H. R. Bode, The role of Hox genes in axial patterning in *Hydra*. *American Zoologist*, **41** (2001), 621–628.

J. R. Finnerty, V. A. Master, S. Irvine *et al.*, Homeobox genes in the Ctenophora: identification of 'paired' type and Hox homologues in the atentaculate ctenophore *Beroe ovata*. *Molecular Marine Biology and Biotechnology*, **5** (1996), 249–258.

J. Q. Henry & M. Q. Martindale, Inductive interactions and embryonic equivalence groups in a basal metazoan, the ctenophore *Mnemiopsis leidyi*. *Evolution and Development*, **6** (2004), 17–24.

Early protostomes

B. C. Boyer, Regulative development in a spiralian embryo as shown by deletion experiments on the Acoel, *Childia*. *Journal of Experimental Zoology*, **176** (1971), 97–106.

J. Q. Henry & B. C. Boyer, The unique developmental program of the acoel flatworm *Neochildia fusca*. *Developmental Biology*, **220** (2000), 285–293.

I. Ruiz-Trillo, M. Riutort, D. T. J. Littlewood *et al.*, Acoel flatworms: earliest extant bilaterian metazoans, not members of Platyhelminthes. *Science*, **283** (1999), 1919–1923.

M.J. Telford, A.E. Lockwood, C. Cartwright-Finch *et al.*, Combined large and small subunit RNA phylogenies support a basal position of the acoelomorph flatworms. *Proceedings of the Royal Society of London B*, **270** (2003), 1077–1083.

J. Baguña & M. Riutort, The dawn of bilaterian animals: the case of acoelomorph flatworms. *BioEssays*, **26** (2004), 1046–1057.

C.E. Cook, E. Jiménez, M. Akam *et al.*, The Hox gene complement of acoel flatworms, a basal bilaterian clade. *Evolution and Development*, **6** (2004), 154–163.

D.H. Erwin & E.H. Davidson, The last common bilaterian ancestor. *Development*, **129** (2002), 3021–3032.

Worms

S. Carranza, J. Baguña & M. Riutort, Are the platyhelminthes a monophyletic primitive group? An assessment using 18S rDNA sequences. *Molecular Biology and Evolution*, **14** (1997), 485–497.

E. Saló, J. Tauler, E. Jiménez *et al.*, Hox and paraHox genes in flatworms: characterisation and expression. *American Zoologist*, **41** (2001), 652–663.

S.A. Maslakova & J.L. Norenburg, Trochophore larva is plesiomorphic for nemerteans: evidence for prototroch in a basal nemertean, *Carinoma tremaphorus* (Paleonemertea). *American Zoologist*, **41** (2001), 1515–1516.

M. Thollesson & J.L. Norenburg, Ribbon worm relationships: a phylogeny of the phylum 'Nemertea'. *Proceedings of the Royal Society of London B*, **270** (2002), 407–415.

J.M. Turbeville, Progress in nemertean biology: development and phylogeny. *Integrated and Comparative Biology*, **42** (2002), 692–703.

M. Shankland & E.C. Seaver, Evolution of the bilaterian body plan: what have we learnt from annelids? *Proceedings of the National Academy of Sciences*, **97** (2000), 4434–4437.

E.M. De Robertis, The ancestry of segmentation. *Nature*, **387** (1997), 25.

Other protostomes

P.N. Lee, P. Callaerts, H.G. de Couet *et al.*, Cephalopod Hox genes and the origin of morphological novelties. *Nature*, **424** (2003), 1061–1065.

R. De Rosa, J.K. Grenier, T. Andreeva *et al.*, Hox genes in brachiopods and priapulids and protostome evolution. *Nature*, **399** (1999), 772–776.

K.M. Halanych, J.D. Bacheller, A.M.A. Aguinaldo *et al.*, Evidence from 18S ribosomal DNA that the lophophorates are protostome animals. *Science*, **267** (1995), 1641–1643.

B.L. Cohen, Monophyly of brachiopods and phoronids: reconciliation of molecular evidence with Linnaean classification (the subphylum Phoroniformea nov.). *Proceedings of the Royal Society of London B*, **267** (2000), 225–331.

M. Kobayashi, H. Furuya & P.W.H. Holland, Dicyemids are higher animals. *Nature*, **401** (1999), 762.

C.L. Anderson, E.U. Canning & B. Okamura, A triploblast origin for Myxozoa? *Nature*, **392** (1998), 346.

A.S. Monteiro, B. Okamura & P.W.H. Holland, Orphan worm finds a home: *Buddenbrockia* is a myxozoan. *Molecular Biology and Evolution*, **19** (2002), 968–971.

D. Papillon, Y. Perez, X. Caubit *et al.*, Hox gene survey in the chaetognath *Spadella cephalaptera*: evolutionary implications. *Development, Genes and Evolution*, **213** (2003), 142–148.

M.J. Telford, Affinity for arrow worms. *Nature*, **431** (2004), 254–256.

Ecdysozoa and the division of the protostomes

W.A. Shear, End of the 'Uniramia' taxon. *Nature*, **359** (1992), 477–478.

M. Averof & S.M. Cohen, Evolutionary origin of insect wings from ancestral gills. *Nature*, **385** (1997), 627–630.

S. Ogg, S. Paradis, S. Gottlieb *et al.*, The Fork head transcription factor DAF-16 transduces insulin-like metabolic and longevity signals in *Caenorhabditis elegans*. *Nature*, **389** (1997), 994–999.

L. Ramskold & X. Hou, New early Cambrian animal and onychophoran affinities of enigmatic metazoans. *Nature*, **351** (1991), 225–228.

A.M.A. Aguinaldo, J.M. Turbeville, L.S. Linford *et al.*, Evidence for a clade of nematodes, arthropods and other moulting animals. *Nature*, **387** (1997), 489–493.

A. Adoutte, G. Balavoine, N. Lartillot *et al.*, The new animal phylogeny: reliability and implications. *Proceedings of the National Academy of Sciences*, **93** (2000), 4453–4456.

A. Graham, Animal phylogeny: root and branch surgery. *Current Biology*, **10** (2000), R36–R38.

W. Arthur, The emerging conceptual framework of evolutionary developmental biology. *Nature*, **415** (2002), 757–764.

Deuterostomes

C.J. Lowe & G.A. Wray, Radical alterations in the roles of homeobox genes during echinoderm evolution. *Nature*, **389** (1997), 718–721.

G.A. Wray & C.J. Lowe, Developmental regulatory genes and echinoderm evolution. *Systematic Biology*, **49** (2000), 28–51.

L.D. Bromham & B.M. Degnan, Hemichordates and deuterostome evolution: robust molecular phylogenetic support for a hemichordate and echinoderm clade. *Evolution and Development*, **1** (1999), 166–171.

C.B. Cameron, Particle retention and flow in the pharynx of the hemichordate worm *Harrimania planktophilus*: the filter feeding pharynx may have evolved before the chordates. *Biological Bulletin*, **202** (2002), 182–200.

K. Tagawa, N. Satoh & T. Humphreys, Molecular studies of hemichordate development: a key to understanding the evolution of bilateral animals and chordates. *Evolution and Development*, **3** (2001), 443–454.

P. Dehal, Y. Satou, R.K. Campbell *et al.*, The draft genome of *Ciona intestinalis*: insights into chordate–vertebrate origins. *Science*, **298** (2002), 2157–2167.

H. Gee, Return of a little squirt. *Nature*, **420** (2002), 755–756.

N.H. Patel, Time, space and genomes. *Nature*, **431** (2004), 28–29.

H.-C. Seo, R.B. Edvardsen, A.D. Maeland *et al.*, Hox cluster disintegration with persistent anteroposterior order of expression in *Oikopleura dioica*. *Nature*, **431** (2004), 67–71.

E.M. De Robertis & Y. Sasai, A common plan for dorsoventral patterning in Bilateria. *Nature*, **380** (1996), 37−40.

S. Bourlat, C. Nielsen, A. E. Lockyer *et al.*, *Xenoturbella* is a deuterostome that eats molluscs. *Nature*, **424** (2003), 925−928.

F. Delsuc, H. Brinkmann, D. Chourrot & H. Philippe, Tunicates and not cephalochordates are the closet living relatives of vertebrates. *Nature*, **439** (2006), 965−968.

Arthropod phylogeny

S.B. Carroll, Homeotic genes and the evolution of arthropods and chordates. *Nature*, **376** (1995), 479−485.

M. Averof & M. Akam, Hox genes and the diversification of insect and crustacean body plans. *Nature*, **376** (1995), 420−423.

M. Averof, Origin of the spider's head. *Nature*, **395** (1998), 436−437.

J.L. Boore, D.V. Lavrov & W.M. Brown, Gene translocation links insects and crustaceans. *Nature*, **392** (1998), 667−668.

M. Akam, Hox genes: from master genes to micromanagers. *Current Biology*, **8** (1998), R676−R678.

M. Akam, Arthropods: developmental diversity within a (super) phylum. *Proceedings of the National Academy of Sciences*, **97** (2000), 4438−4441.

M. Levine, How insects lose their limbs. *Nature*, **415** (2002), 848−849.

T.R. Galant & S.B. Carroll, Evolution of a transcriptional repression domain in an insect Hox protein. *Nature*, **415** (2002), 910−913.

C.E. Cook, M.L. Smith, M.J. Telford *et al.*, Hox genes and the phylogeny of arthropods. *Current Biology*, **11** (2001), 759−763.

F. Nardi, G. Spinsanti, J.L. Boore *et al.*, Hexapod origins: monophyletic or paraphyletic? *Science*, **299** (2003), 1887−1889.

Homology

D. Tautz, Debatable homologies. *Nature*, **395** (1998), 17−19.

J.A. Bolker & R.A. Raff, Developmental genetics and traditional homology. *BioEssays*, **18** (1996), 489−494.

B.K. Hall (ed.), *Homology: the Hierarchical Basis of Comparative Biology* (San Diego: Academic Press, 1994).

B.K. Hall, Descent with modification: the unity underlying homology as seen through an analysis of development and evolution. *Biological Reviews*, **78** (2003), 409−433.

J.D. McGhee, Homologous tails? Or tales of homology? *BioEssays*, **22** (2000), 781−785.

R. Quiring, U. Waldorf, U. Kloter *et al.*, Homology of the eyeless gene of *Drosophila* to the small eye gene in mice and aniridia in humans. *Science*, **265** (1994), 785−789.

W.J. Gehring & K. Ikeo, *Pax 6*: mastering eye morphogenesis and eye evolution. *Trends in Genetics*, **15** (1999), 371−381.

S.I. Tomarev, P. Callaerts, L. Koss *et al.*, Squid *Pax-6* and eye development. *Proceedings of the National Academy of Sciences*, **94** (1997), 2421−2426.

Glossary

Words in *italic* have their own entries in the Glossary.

Abdomen The third and most posterior division of the body of an insect, applied also to describe crustaceans.

Abductor When contracted an abductor muscle separates two hard structures, or draws a limb away from the midline of the animal (contrast *adductor*).

Aboral Away from the mouth, e.g. the upper surface of a starfish or a sea urchin.

Actin Globular protein which readily polymerises into a fibrous form and is an important component of *microfilaments*. Actin constitutes the thinner fibrils involved in muscle contraction (Box 5.1).

Action potential The nerve impulse passing along an *axon*.

Adaptation Any change in structure or function that makes part or the whole of an organism better suited to its function or to its environment. It is the result of *natural selection*.

Adaptive radiation The divergent evolution of different species from a common ancestor, due to *natural selection*.

Adductor When contracted an adductor muscle pulls together two hard structures (e.g. the two halves of a bivalve shell), or draws a limb towards the midline of the animal.

Afferent Carrying blood or nerve impulses from the outer parts of the body towards the centre (contrast *efferent*).

Alimentary canal Food cavity (gut) which in most animals leads from the mouth to the anus.

Allele (short for **allelomorph**) Alternative forms of *genes* occupying the same locus (position) on a pair of *chromosomes*.

Alternation of generations Alternation between a *haploid* and a *diploid* generation within an organism's life cycle (does not apply to cnidarian *polyps* and *medusae*, where the chromosomes are identical in number and kind in the two forms). Most animals confine the haploid stage to the *gametes*.

Ambulacral Concerned with walking; applied to appendages or to grooves bearing *tube feet* in echinoderms.

Ammonia NH_3 (ion NH_4^+), the primary product of nitrogenous metabolism. Released directly into the environment by most marine organisms, but it is very toxic. Where water is not freely available, energy must be spent synthesising ammonia into less toxic forms (*urea*, *uric acid* etc.) for excretion.

Amoebocyte   A cell that moves by *amoeboid* movement, e.g. in a sponge.

Amoeboid   Any movement resembling that of the protistan, *Amoeba*, characterised by changes in shape of the moving cell.

Ampulla   A small vesicle containing liquid, e.g. those associated with the *tube feet* in many echinoderms.

Anaerobic (or **Anoxic**)   Lacking oxygen. Anaerobic respiration occurs in many invertebrates: carbohydrates are incompletely broken down, producing lactic acid rather than carbon dioxide and water, and making less energy available.

Animal pole   The end ('pole') of an egg containing the *nucleus*; usually uppermost, since any heavy yolk concentrates in the *vegetal* end.

Antenna   Anterior sensory structure, typically a long thin process in an arthropod, sensitive to mechanical and chemical stimuli.

Apodeme   A piece of the exoskeleton of an arthropod, folded internally and attaching muscle.

Aquatic animal   One living in water.

Aristotle's lantern   The feeding structure of a sea urchin.

Asexual   Reproduction, usually by fission or budding, that does not involve either sexual fusion or *meiotic* nuclear division.

Atrium   Chamber surrounding the *pharynx* in some invertebrate chordates, receiving water through the pharyngeal gill slits and opening to the exterior by a more posterior atriopore.

Autotomy   The self-induced dropping off of part of an organism (often a limb), usually in response to threat from a predator.

Axon   The main projection from a nerve cell, along which nerve impulses pass.

Bilateral symmetry   When only one plane perpendicular to the oral/*aboral* axis can divide the animal into two mirror-image halves (contrast *radial symmetry*).

Bilateria (adj. **bilaterian**)   Group name for *triploblastic* animals, most of which have *bilateral symmetry*.

Biramous   Having two branches.

Blastocoel   The primary cavity, appearing at the end of *cleavage* in the developing embryo.

Blastomere   Cell within a *blastula* in early *cleavage*.

Blastopore   The first opening of the embryonic gut. It becomes the mouth in *protostomes* and the anus in *deuterostomes*.

Blastula   Hollow ball of cells formed by the *cleavage* of the developing egg.

Body cavity   Fluid-containing cavity between the gut and the outer surface of an animal, containing internal organs.

Botryoidal tissue   Tissue consisting of bunches of cells and containing a dark pigment, in leeches. Derived from *parenchyma*, it invades the *coelom*, leaving only a few channels serving for fluid transport.

Brain   Aggregation of *nerve cells*, usually anterior, which acts as a sensory/motor exchange (in simple animals) or a controlling nerve centre for the whole body.

Buccal cavity   Space just inside the mouth, at the anterior end of the alimentary canal.

Buoyancy Neutral buoyancy is achieved when a body floats at a constant level in the water. Buoyancy is said to be negative if the body is denser than the water and therefore tends to sink.

Byssus Adhesive threads secreted by the foot of a bivalve mollusc, attaching the animal to the substratum.

Caecum (pl. **caeca**) Pouch-like diverticulum of an animal's gut, usually functioning in digestion. In many herbivorous animals the caecum houses *symbiotic* microorganisms able to digest cellulose.

Carapace The protective cuticular 'shell' of many crustaceans, overlying the main body cuticle. Usually anterior only.

Catch muscle Misleading name given to muscles such as bivalve *adductors* which can maintain low tension for a long time; misleading because the muscle is not locked in position but requires a regular small nerve input and contraction is actively, if very economically, maintained.

Caudal Tail.

Cephalothorax Combined head and thorax of many crustaceans.

Cercus (pl. **cerci**) One of a pair of appendages on the last abdominal segment of an insect.

Chaeta (pl. **chaetae**) Small stiff chitinous bristles of annelids and other worms (also called Setae).

Chela (pl. **chelae**) Claws, pincers.

Chelicera (pl. **chelicerae**) One of the first pair of appendages on the *prosoma* of a chelicerate. Often functions to seize and grasp food.

Chemosynthesis Manufacture of organic food from its inorganic constituents using energy from chemical processes rather than (as in photosynthesis) light. Largely confined to bacteria, but many animals rely on *symbiotic* chemosynthetic bacteria.

Chitin Nitrogen-containing polysaccharide which is a major component of the *cuticle* of many animals, notably arthropods. Chitin is soft, but combined with cross-linked protein chains it makes a hard and light covering (see *tanned protein*).

Chlorocruorin Rare iron-containing *respiratory pigment* of some annelids.

Choanocyte Characteristic cell constituting the inner layer of a sponge, consisting of a nucleated cell body in which a *flagellum* is rooted and a collar made of *microvilli* where food particles are ingested.

Chromatophores Cells just under the skin of some animals bearing pigment granules. Aggregation or dispersal of chromatophores changes the colour of the animal.

Chromosome Thread-like structure in the cell nucleus carrying the *genes*. Chromosomes are made of *DNA* and protein. They may be visible by light microscopy at various stages in cell division.

Cilium (pl. **cilia**) Short (usually up to 10 μm long) hair-like thread borne on a cell, usually present in large numbers. Ciliary beat generates water currents which may move fluid through the animal or the animal through the water.

Cladistic analysis Aims to trace ancestry by constructing *cladograms* based on combination of organisms sharing the same difference from the primitive condition, aiming to detect *monophyletic groups*.

Cladogram A series of dichotomous branches depicting a hierarchy of recency of common ancestry, based on the presence of *shared derived characters*.

Cleavage The earliest divisions of the fertilised egg, proceeding by *DNA* replication and (usually) subdivision of the cytoplasm.

Clitellum The 'saddle' familiar in an earthworm and occurring in all oligochaetes and leeches. It collects the gametes and wriggles over the worm's head, forming a cocoon where fertilisation and development occur.

Clone A group of genetically identical cells or organisms derived asexually from a single founder cell.

Cnidae See *nematocysts*.

Coelenteron The water-containing inner cavity of cnidarians, open to the outside world. It is in the position of a gut and is not technically a '*body cavity*', although it may serve as a *hydrostatic skeleton*.

Coelom *Body cavity* entirely surrounded by *mesoderm*: in section it appears between two layers of mesoderm.

Collagen A fibrous protein which forms a large part of animal *connective tissue*. Polypeptide chains of the protein form helical coils bound together to form fibrils which have great strength and little elasticity.

Commensal Sharing the food of another animal with which it lives. Often one member of the pair benefits and the other is not strongly affected.

Commitment (or **determination**) The limitation of the developmental capacity which directs the cell towards a particular pathway, the prelude to *differentiation*.

Compound eye Single eye formed of many thousands of optical units, *ommatidia*, each with its own lens and light-sensitive pigment. Occurs in arthropods and some other animals.

Connective tissue Animal tissue serving for support and packing; characterised by a large quantity of ground substance (matrix) usually containing fibres, and relatively few cells.

Contractile vacuole Structure within a cell that expands as it fills with water and contracts as it empties its contents to the exterior. A means of water regulation found in *Amoeba* and freshwater sponges.

Convergent evolution Occurs when resemblances are due to common adaptations to a particular function or way of life, rather than due to close common ancestry. Cannot be distinguished satisfactorily from *parallel evolution*, and these terms are best regarded as synonymous.

Crossing over The exchange of parts of *homologous chromosomes* when they come together at *meiosis* (see Figure 1.2).

Cryptozoic 'Hiding animals' inhabiting dark secluded habitats such as crevices.

Ctenidium (pl. **ctenidia**) A molluscan gill that is suspended in the *mantle cavity*, used for respiration or trapping of small food particles. Primitively it consists of rows of filaments borne on a common axis.

Cuticle Outermost protective structure of an animal, often made partly of horny material such as *collagen*, secreted by underlying cells. Flexibility may be combined with hard regions, as in the chitinous cuticle of arthropods which forms an exoskeleton needing to be moulted as the animal grows.

Cytoplasm All the cell contents bound within an outer cell membrane, other than the *nucleus*.

Dendrite Projection of nerve cell bodies which make contact with other nerve cell projections.

Depolarisation Applied to nerve: a decrease in the *resting potential* across the nerve membrane.

Derived character Secondary or specialised, as opposed to a *primitive* character.

Determination See *commitment*.

Detritus Particle of organic matter in water, often decomposed, floating or associated with the substratum.

Deuterostome The group of animals where the embryonic blastopore typically gives rise to the anus and the mouth develops secondarily elsewhere (e.g. echinoderms and chordates). Typically deuterostomes share other characters of early development (contrast *protostome*; see Box 5.2).

Diapause A period of arrested development and low oxygen consumption in an insect life cycle, whose onset is determined by some event in past time (often a change in day length).

Dichotomy Division into two.

Differentiation The appearance of differences between cells during development; preceded by *commitment*.

Diffusion Movement of a substance from a region of higher concentration to a region of lower concentration of that substance, freely into the surroundings or across a permeable or semi-permeable membrane. Diffusion is passive in that its occurrence does not require energy input.

Diploblastic Having two cell layers only, the *ectoderm* and the *endoderm*, as in Cnidaria.

Diploid Having the full number of *chromosomes* normal for the species. The chromosomes are present in pairs, one being derived from each parent if reproduction is sexual (contrast *haploid*).

Direct development Hatching from the egg in the adult form, without an intermediate larval stage.

Distal The end of a limb or other structure that is furthest from the centre of the body (or from its point of attachment, if the animal is *sessile* (contrast *proximal*).

DNA (deoxyribonucleic acid) The genetic material of all living organisms except some viruses. A major constituent of the *chromosomes* in the *nucleus* and also occurring in cytoplasmic organelles such as *mitochondria*.

Dominant Genes expressed when present either as both members of an *allelomorphic* pair or as only one member of that pair (contrast *recessive*).

Dorsal Upper surface or back of an animal.

Double helix The genetic material: a spiral ladder-shaped molecule of *DNA* made by the cross-linking of pairs of bases in two chains of *nucleotides*.

Ecdysis The splitting and discarding of the old *cuticle* which occurs when an arthropod moults.

Ecdysone Hormone implicated in the moulting of arthropods and having various functions in other phyla.

Ecology (original scientific use) The interaction of organisms with each other and with the environment.

Ectoderm Outermost cell layer of an animal's body: may form *epidermis* (skin) and its outgrowths, *nephridia*, nervous tissue.

Efferent Carrying blood or nerve impulses away from the centre of the body (contrast *afferent*).

Efficiency (1) Physical efficiency is greatest for a structure or process requiring the minimum energy input for a given output. (2) Biological efficiency is greatest for a structure or process giving most selective advantage to an animal, which is frequently the same as (1) but not necessarily so.

Endocuticle Soft innermost layer of arthropod *cuticle*, containing chitin not hardened by cross-linked proteins. Reabsorbed into the body at moulting.

Endoderm Innermost of the cell layers of a *diploblastic* or *triploblastic* body. Gives rise to the lining of the gut (alimentary canal) and its associated glands.

Endostyle Ciliated mid-ventral groove in the *pharynx* of some invertebrate chordates, *mucus* secreting and able to take up iodine. Forerunner of the vertebrate thyroid gland.

Enterocoel *Coelom* formed by out-pouching from the embryonic gut, as in echinoderms and invertebrate chordates.

Epiboly Method of *gastrulation*, by the spreading of a sheet of outer cells around an inner layer, which hollows out to form the gut.

Epicuticle Structure overlying the *exocuticle* of arthropods, containing various amounts of wax.

Epidermis Outermost living layer of many animals, characteristically a large number of similar cells in close contact.

Eukaryote Organism with cells having a distinct nucleus surrounded by a double membrane (unlike bacteria and Archaea); nearly always with *mitochondria*.

Euryhaline Aquatic animals able to tolerate a wide range of salinity, whether as *osmoconformers* or as *osmoregulators*.

Evolution Descent with modification, the process by which the diversity of organisms arose from the earliest forms of life. Evolution is the result of *natural selection*. It has been called the modification of development by ecology.

Excretion Removal of metabolic waste (i.e. waste materials formed by the body's own activity, not including undigested food). The term may be extended to all regulation of the internal body fluids, i.e. ion and water regulation, since the same organs are often involved.

Exocuticle Layer of the arthropod exoskeleton immediately under the *epicuticle*, containing *chitin* and hardened by cross-linked *tanned proteins*. Shed at moulting.

Exoskeleton Hard skeleton on the outer surface of an animal's body, such as the *cuticle* of an arthropod.

Flagellum (pl. flagella) Relatively long (*c.* 150 µm) fine whiplike projection from the surface of a cell, e.g. sperm, *flame cell*, sponge *choanocyte*. Like cilia, the beat causes movement in the surrounding fluid; unlike cilia, flagella are long enough to have several waves at the same time.

Flame cell Blind-ending *protonephridial* tube with a beating flagellar 'flame' drawing in water and solutes from the surroundings through the walls of the tube.

Fossils Remnants (dating back to before the last Ice Age) of once-living animals, preserved in rocks or (less often) sediments, amber, ice etc. Hard parts may be preserved, or footprints, tracks etc. (see *trace fossils*).

Free-living (1) Not *parasitic*. (2) Not confined to one spot but capable of locomotion in the environment.

Gamete The *haploid* unit (usually differentiated as egg or sperm) that fuses with another at sexual reproduction.

Ganglion (pl. **ganglia**) Collection of *nerve cell* bodies, as opposed to *axons*.

Gastrodermis Alternative description of an *endodermal* layer, e.g. in Cnidaria.

Gastrulation Series of cell movements in early development that brings the cells that will form internal organs to the inside of the developing embryo. Gastrulation converts a ball or sheet of cells into a multilayered structure within which the future germ layers take their correct relative positions.

Gel electrophoresis Separation in an electric field of colloidal substances such as gels: a method of identifying the components of a mixture by the different rates at which they move between two applied electrodes.

Gene Unit of heredity that passes from parent to offspring. Part of a molecule of *DNA* borne on a *chromosome*, a gene consists of a sequence of nucleotide bases coding for all or part of a protein. It is the shortest length of a DNA molecule that can undergo *mutation* or *recombination*.

Genetic drift Random fluctuations in *gene* frequency which are not due to *natural selection*, as when a small population within a species is isolated and *evolution* depends on the genes that happen to be present.

Genome Sum total of the *genes* contained within an organism.

Genotype Genetic constitution of an organism, as opposed to the *phenotype*.

Genus (pl. **genera**) A group used in classification containing one species with unusual characters or (commonly) many species believed to be closely related. It is a working hypothesis, necessarily imprecise but used as an estimation of relatedness.

Germ plasm The part of an organism specialised for sexual reproduction, i.e. for transmission of hereditary characters down the generations.

Glycogen Large insoluble carbohydrate storage molecule: 'animal starch'.

Gnathobase Process arising from an arthropod appendage, used for manipulating hard food.

Gonad A structure producing *gametes*, e.g. an ovary producing eggs and a testis producing sperm.

Guanine Insoluble purine which is the end point of nitrogenous metabolism in spiders. Can be excreted with very little loss of water.

Haemal Concerning blood. Misapplied to one of the *coeloms* of an echinoderm.

Haemerythrin Rare iron-containing *respiratory pigment* of some annelids.

Haemocoel Blood-filled *body cavity* derived from the *blastocoel*, in arthropods and molluscs.

Haemocyanin Copper-containing *respiratory pigment* in many crustaceans and molluscs.

Haemoglobin Iron-containing *respiratory pigment* occurring in almost every phylum.

Haltere Modified hind wing or 'balancer' of a dipteran fly, in the form of a stick with a heavy head, beating with the wings. Strain receptors at the haltere base inform the fly of any irregularities needing to be corrected.

Haploid Having half the number of *chromosomes* normal for the particular species, as when *gametes* are formed, each with a single chromosome set.

Hermaphrodite Having both male and female sex organs in one individual, which may be cross-fertilised, or more rarely self-fertilised.

Homeobox Sequence of 180 *nucleotide* bases which code for a sequence of amino acids, the *homeodomain*. The homeobox has been strongly conserved during evolution: genes containing it regulate development. Hox genes are examples of homeobox-containing genes.

Homeodomain Conserved sequence of 60 amino acids coded for by the *homeobox*. Interaction between the homeodomain and target *DNA* sequences may determine whether or not the target *genes* are expressed.

Homeotic Describes *genes* that select between alternative pathways in early development. *Mutations* in homeotic genes result in one part of the body 'becoming like' another part, as when *Drosophila* mutants form a leg in place of an *antenna*.

Homology (adj. **homologous**) Morphological similarity due to a common evolutionary origin. This is a fundamental concept in tracing evolutionary relationships, yet there is no simple universally accepted precise definition (for example, as to whether a common developmental path should be invoked). See Chapter 20.

Homologous chromosomes In this specialised usage 'homologous' approximates to 'similar'. Pairs of structurally similar *chromosomes* can be identified at the outset of *meiosis*; one member of each pair comes from each parent, and both members of a pair have the same pattern and positions of genes but the *alleles* may be different.

Hormone Substance made by a gland or (as often in invertebrates) a specialised nerve cell (see *neurosecretion*) and released into the blood, or into other tissues or fluids. The hormone travels to a distant site where it exerts its effect, often by regulating cell performance. Hormones, in contrast to nerves, may provide an alternative slower and longer-lasting method of coordinating the body.

Hox genes *Homeobox*-containing highly conserved genes that determine the relative position of structures along the main body axis. They initiate a hierarchy of regulatory genes.

Hydrostatic skeleton A fluid-filled cavity in soft-bodied animals, against which muscles contract.

Hyperpolarisation Increase in the *resting potential* across a nerve cell membrane.

Hypodermis Layer of living cells underlying the thick outer *cuticle* in a nematode. Equivalent to the *epidermis* of other animals with cuticles.

Imaginal disc Group of undifferentiated cells in an insect larva, inactive and set aside, ready to develop into a specific adult structure at *metamorphosis*.

Imago (adj. **imaginal**) Adult insect.

Integument Outer layer or skin (but see *tegument*).

Intercellular Between cells (contrast *intracellular*).

Interneuron Nerve cell in a circuit between sensory input and motor output, which it may control by changes in polarisation or depolarisation. Mostly interneurons do not give rise to *action potentials*. They are for example the sites of organisation of flight patterns in insects.

Interstitial (1) Unspecialised cells in Cnidaria and a few worms, able to move between other cells and capable of differentiating into them. (2) Ecological microhabitat of spaces in between particles of soil or (most frequently) marine sand grains.

Intracellular Within a cell (contrast *intercellular*).

Introvert Front end of body of some worms which can be everted or retracted into the body.

Invagination Tucking in: applies to a common method of *gastrulation*, by the infolding of a sheet of cells from a point on the *blastula* surface.

Ionic regulation Selecting and keeping particular ions within the body at concentrations different from those of the surrounding water, even though osmotic regulation may not occur.

Isometric The sliding and cross-linking of *myofilaments* in muscle contraction to build up tension with minimal shortening of the muscle as a whole (contrast *isotonic*).

Isotonic (1) Muscle contraction by sliding of *myofilaments* to shorten the muscle with a minimum change in tension but considerable change in length (contrast *isometric*). (2) The absence of net ion movement between two solutions separated by a semi permeable membrane (often used where 'iso-osmotic', equivalence of ion concentration, would be a more accurate description).

Junction potential Localised electrical changes under the endings of an arthropod motor nerve on the surface of a muscle, not normally propagated over the whole surface.

Labium The combined second *maxillae* (the most posterior pair of head appendages) of some insects and myriapods, forming a 'lower lip'.

Larva The pre-adult form in which some animals hatch from the egg (see Box 17.1 for the difficulties of fuller definition).

Lattice Spirally coiled thread of *collagen*, incompressible in length, situated in or under the outer layer of a soft-bodied invertebrate and wound round the body in a helix. The extension and contraction of this helix is an essential element in the control of muscle contraction in animals with *hydrostatic skeletons*.

Linkage Describes the repeated absence of separation between parts of a *chromosome* at *crossing over* during nuclear division. The closer *genes* are along a chromosome, the more likely they are to be linked: this was the original basis for all gene mapping.

Lophophore A circular or horseshoe-shaped fold of the body wall encircling the mouth (but not the anus) bearing hollow tentacles, each containing a branch of the *mesocoel*. Cilia beat food-bearing water currents from the tip of each tentacle to the base, where the water leaves the animal.

Lung books (or **book lungs**) Respiratory organs of arachnids consisting of 'books' of blood containing leaflets with air channels in between.

Madreporite Echinoderm sieve plate connecting the *water vascular system* with the sea water outside.

Malpighian tubules *Excretory* organs in insects and some other arthropods. Long narrow tubes growing out from the junction of the mid and hind gut, ending blindly in the blood-filled *haemocoel*. Active ion transport from the blood to the tubule lumen draws other ions, water and excretory products into the tubule.

Mandibles Paired appendages just behind the mouth in many insects and crustaceans; hard structure used for seizing and biting food.

Mantle cavity Outside the main body wall of a mollusc, covered wholly or partly by the mantle (a fold of skin), the mantle cavity usually contains *ctenidia* suspended in ambient water. In a pulmonate terrestrial snail or slug it contains air instead of water.

Marine In the sea.

Maxilla (pl. **maxillae**) One of a pair of feeding appendages behind the mouth (in front of the *mandibles*) in most insects and crustaceans.

Maxillipedes Thoracic appendages of crustaceans which assist the *maxillae* in food collection.

Medusa The 'jellyfish' form of individual in Cnidaria, typically umbrella- or bell-shaped, freely floating or swimming, mouth downwards.

Meiosis 'Reduction division' of the cell *nucleus*, producing daughter cells with half the number of *chromosomes* present in the parent (see Figure 1.2).

Mesocoel Coelomic cavity, the middle of the three divisions of the coelom typically occurring in *deuterostomes*. Also provides the hydraulic system of a *lophophore*.

Mesoderm Middle of the three cell layers of a *triploblastic* animal. The site of muscle, sex organs and many other structures.

Mesoglea Jelly-like material between the two cell layers of a cnidarian, containing cells and fibres derived from those layers. Much expanded in jellyfish.

Mesohyl Jelly-like substance between the outer cells and the *choanocytes* in a sponge, containing wandering *amoebocytes*.

Messenger RNA (mRNA) Small molecules of *RNA* transcribed from the *DNA* of the genes and passing out of the *nucleus* into the *cytoplasm* and to the *ribosomes*. Characteristically short-lived except in egg cells, where long–lasting mRNA provides initial maternal instructions for development.

Metabolism The sum total of chemical reactions within the body of an organism.

Metabolite Product of *metabolism*.

Metachronal rhythm Pattern of movement in which each unit (*cilium* or muscle block) contracts just before or just after its neighbour.

Metacoel Most posterior coelomic cavity in a *deuterostome*.

Metameric segmentation Serial repetition of equivalent parts in an animal.

Metamorphosis Pronounced change in form during the animal life cycle.

Metanephridium *Nephridium* opening at both ends, being an excretory channel between the *body cavity* and the outside. Occurring in many worm phyla; often segmentally arranged, as in the excretory organs of most annelids.

Microfilaments Contain the protein *actin* and are usually abundant under the cell membrane. Important in cell motility.

Microtubules Small tubular structures (15 to 25 nanometers) made of subunits of the protein tubulin, occurring in large numbers freely in the cytoplasm and in cell organelles such as *cilia* and *flagella*. They appear to be important in transport of materials for cell motility.

Microvilli Small finger-like projections of the outer cell membrane, filled with *mitochondria* and occurring in large numbers at any site of active absorption, *secretion* or sensory perception.

Mitochondrion (pl. **mitochondria**) The organelle within the cytoplasm that bears the enzymes that carry out aerobic respiration. Very numerous in cells with high metabolic activity. Mitochondria contain their own *DNA*, different from that of the cell nucleus and in the circular form found in bacteria (a pointer to the *symbiotic* origin of metazoan cells). Mitochondria are transmitted down the generations, normally only in the female line, and their DNA has become very important in tracing evolutionary relationships.

Mitosis Division of the *nucleus* producing two daughter nuclei with *chromosomes* identical in number and kind to those of the parent.

Molecular clock Said to be regular if molecular evolution (i.e. substitutions in the *nucleotide* bases of *DNA*) proceed at a constant rate, so that the degree of difference between the DNA of two species can be used to measure time elapsed since their evolutionary divergence. The difficulty is that even if this form of *neutral evolution* does occur, the rate of molecular change is known not to be constant in different species in different conditions.

Monophyletic A group derived from a single ancestor, including that ancestor and all of its descendants. *Cladistic analysis* aims to reveal monophyletic groups.

Morphogen Diffusing substance conferring the *polarity* of an organism and/or initiating certain kinds of development which do not occur without it.

Mosaic Made of many parts that together make a whole, applied (1) to the development of *spirally cleaving* eggs where *commitment* occurs early and (2) to the *compound eyes* of arthropods. Neither of these attributions is very accurate.

Mucus (adj. **mucous**) Polysaccharide bound to a protein, secreted by most animals in most phyla and possessing very important properties: it can be very liquid, gelatinous, or (with some chemical change) adhesive. Its uses include lubrication, protection against desiccation, binding of food particles and adherence to the substratum.

Multicellular Organisms (or parts of them) made of many cells (e.g. all Metazoa).

Mutation Change in the structure of a *gene* or *chromosome* set, typically random in appearance. Mutations may lead to heritable variations, the raw material on which *natural selection* acts, causing *evolution*.

Myofilaments *Actin* and *myosin* fibrils which slide between each other when muscle contracts.

Myogenic Originating in or produced by muscle cells, e.g. the rapid oscillation of the 'asynchronous' 'fibrillar' flight muscles of some small insects.

Myoglobin Simple form of *haemoglobin* present in various tissues; may store or facilitate the diffusion of oxygen.

Myosin Protein with a globular head and a rod-like tail; the main constituent of muscle fibrils, constituting the thicker of the *myofilaments*.

Natural selection Process, discovered by Darwin, that brings about *evolution*: the differential reproduction of individuals with attributes favourable to their particular habitat or way of life, imprecisely formulated as 'the survival of the fittest'. Results in the *adaptation* of structure to function and organism to environment.

Nauplius Characteristically the first *larva* of crustaceans, having simple eyes and three pairs of appendages.

Nematocysts (**cnidae**) Stinging cells unique to the phylum Cnidaria. Reception of a stimulus and response (by the ejection of a stinging thread) is carried out by the single cell, which may act without nervous stimulation.

Neoteny *Paedomorphosis* by delay in the development of adult characters so that the animal reproduces while still in the *larval* form (contrast *progenesis*).

Nephridium (pl. **nephridia**) Excretory organ made in one of two ways: either by in-tucking of an epidermal tube which remains closed at its inner end (as in *protonephridia* such as *flame cells*), or by a mesodermal duct, open at both ends, leading from the *coelom* to the outside world (as in the *metanephridial* or 'segmental organs' of many coelomate animals).

Nerve cell See *neuron*.

Nerve net Network of nerve cells connected by non-polarised junctions and able to conduct in all directions. Occurs in various worms and in Cnidaria and Echinodermata, where there is no controlling brain.

Neuron (neurone) Unit of conduction by the nervous system, consisting of a cell body with processes (*dendrites*) and usually one long process, the *axon*, conducting nerve impulses.

Neurosecretion Production of *hormones* by specialised cells in the nervous system. Neurosecretory cells may be recognisable by large rounded nuclei in the cell body: the ends of the axons secrete hormones acting at a distance rather than nerve transmitters that bridge a *synapse*.

Neutral evolution Descent with modification not due to the action of *natural selection*, causing accumulation in the *genome* of selectively neutral *mutations*. The extent to which it occurs continues to be controversial.

Niche An ecological niche defines not only where an organism lives but also its type of food and general requirements.

Notochord Dorsal stiffening rod characteristic of Chordata, made of closely fitting cells with hydrostatic pressure due to intracellular *vacuoles* and covered in a fibrous sheath.

Nucleolus Region of the cell nucleus especially rich in *RNA* and *protein*. The site of origin of *ribosomes*.

Nucleotide Unit of structure of nucleic acids, *DNA* and *RNA*, consisting of a sugar (deoxyribose or ribose) with one hydroxyl (OH) group bound to a purine or pyramidine nitrogenous base.

Nucleus (pl. **nuclei**) Cell organelle containing the *chromosomes*, hence the *genes*, and bounded by a nuclear membrane. Present in all cells, at their origin at least.

Numerical taxonomy See *phenetic taxonomy*.

Nymph Old-fashioned term for the *larva* of an exopterygote insect.

Oesophagus Part of the gut between the *buccal cavity* and the stomach. There may or may not be a distinct *pharynx* between the buccal cavity and the oesophagus.

Ommatidium (pl. **ommatidia**) One of the light-sensitive units in an arthropod *compound eye*, consisting of a cuticular lens and a number of light-detecting retinula cells, the whole being surrounded by pigment.

Ontogeny Development of an individual from an egg to an adult.

Opisthosoma Posterior division of the body of a chelicerate, without locomotor appendages but with gills in aquatic forms.

Osculum Large hole in a sponge through which a jet of water is ejected.

Osmoconformer Animal (such as most marine invertebrates) whose internal body fluids have approximately the same osmotic potential as the surrounding water.

Osmole The osmotic effect of a solute when the molecular weight of the solute is dissolved in a litre of water. For sea water the total osmotic effect of the solutes is 1 osmole (1000 milliosmoles, mosm).

Osmoregulation Maintenance of the internal body fluids at a concentration (and hence osmotic potential) different from that of the surrounding water.

Ossicle   Skeletal unit of an echinoderm: a calcareous plate separated by living tissue from its neighbours, at least while it develops.

Ostium (pl. **ostia**)   (1) One of many small holes perforating the body of a sponge, through which water enters.   (2) Holes in an arthropod heart through which blood enters from the *haemocoel*.

Outgroup   Group of species used to define the root of a *cladogram*, chosen as evolutionarily close to the group in the cladogram but not a part of it.

Ovoviviparity   Embryos are retained in yolky eggs within the mother's body while they develop, but are not fed by the mother (contrast *viviparity*).

Oxygen tension   Measure (by pressure, expressed as mm mercury in a barometer) of the availability of oxygen in a solution.

Paedomorphosis   Development to reproductive maturity of the juvenile form of an animal, either by *neoteny* or by *progenesis*.

Papilla   A cone-shaped projection from the surface of an organ or an organism.

Papula   Respiratory projections from the echinoderm water vascular system.

Parallel evolution   See *convergent evolution*.

Paraphyletic   Assembly including some but not all members of a monophyletic group.

Parapodium (pl. **parapodia**)   Paddle-like lobes projecting from the sides of the segments of polychaete worms. Used mainly for moving water, sometimes involving locomotion.

Parasite   Specialised form of predator which does not kill its prey: it lives inside or attached to the outer surface of a host (of another species) and feeds on the host's living body. The host derives no benefit from the association.

Parenchyma   Unspecialised middle layer of some simple *triploblastic* animals. Characterised by few cells, assorted fibres and intercellular spaces filled with liquid.

Parsimony   Originally defined in the application of 'Occam's razor' (the acceptance of simple explanations in preference to more complicated ones), parsimony has been annexed by cladists to describe choice of the one of many possible *cladograms* postulating the smallest number of evolutionary steps, i.e. the minimum need to postulate *convergence*. This process minimises the perception of convergence and may or may not be correct.

Parthenogenesis   Development from an unfertilised egg in a female animal.

Pedicellarium (pl. **pedicellaria**)   Small stalked pincers on the surface of an echinoderm, used for grasping and defence.

Pedipalp   One of the second pair of appendages on the *prosoma* of a chelicerate. Functions vary from seizing and killing prey to use as walking or digging legs.

Pelagic   Organisms swimming or drifting in the main body of the sea or a lake, whether *planktonic* or in deeper waters.

Pericardium   Membrane round the heart. The pericardial cavity is a *body cavity* containing the heart.

Peristalsis   Movement of part or the whole of an organism by waves of alternate contraction and extension, e.g. movement of food through the intestine or of an earthworm through soil, achieved by circular and longitudinal muscles contracting in turn.

Pharynx (adj. **pharyngeal**)   The part of the alimentary canal behind the *buccal cavity* and anterior to the *oesophagus*.

Phenetic taxonomy (numerical taxonomy) Proceeds by putting together those organisms with the greatest number of common characters.

Phenotype Manifested attributes of an organism. Due primarily to gene activity, but contrast *genotype*.

Pheromone *Hormone* acting outside the body.

Phylogenetic The relationship between organisms and groups of organisms that is due to their evolutionary history. Description of *taxonomy* based on cladistics.

Phylogeny The pattern of evolutionary relationships.

Phylum (pl. **phyla**) Group of animals with a common body plan.

Pinacocyte Outer cell of a sponge.

Plankton Surface waters of the sea or a freshwater lake. Applied to small animals and plants drifting or swimming in surface waters; they provide a very important source of food.

Pluteus Echinoderm larva characteristic of echinoids and ophiuroids.

Polarisation (of a nerve membrane) Presence of the *resting potential*.

Polarity (1) Direction or order of succession in time or space; hence (2) Definition of the axis of an organism or part of an organism.

Polymorphism Different forms occurring within one species.

Polyp One form of individual within the phylum Cnidaria, typically fixed at one end and bearing tentacles round the mouth at the other end.

Polyphyletic Assemblage of animals with different ancestral lineages.

Predator Animal which kills and feeds on another animal.

Primitive Earliest, not necessarily simplest, example of a character or organism; may or may not be preserved to the present time.

Proboscis (pl. **proboscides**) Any tubular projection at the anterior end of an animal, whether it is part of the gut or separate from it (as characterises nemertines) or is the trunk of an elephant.

Progenesis *Paedomorphosis* by acceleration of development of the gonads; the animal reproduces in the larval form and the generation time is shortened.

Proprioceptor Sensory receptor responding to changes within the body, e.g. stretch receptors in muscle.

Prosoma Anterior division of the body in chelicerates.

Protandrous *Hermaphrodite* in which the male organs develop and function first.

Protein Large group of complex nitrogen-containing organic compounds, made up of polypeptides consisting of strings of amino acids. The primary constituent of living matter, including enzymes.

Protocoel Most anterior *coelomic* cavity of a *deuterostome*.

Protogynous *Hermaphrodites* in which the female organs develop and function before the male ones.

Protonephridium *Nephridium* with a closed end, as in *flame cells*. Occurs in platyhelminths, nemertines, a few polychaetes and larvae.

Protostome *Triploblastic* animals in which the *blastopore* usually becomes the mouth, with other distinctive features of early development (contrast *deuterostome*).

Proximal The end of a limb or other structure that is nearest to the centre of the body or, in a sessile animal, its point of attachment (contrast *distal*).

Pseudocoel Obsolete term for an ill-defined *body cavity*.

Pupa Stage between larva and adult in an endopterygote insect. Known as a resting stage, but is a period of great internal reorganisation as larval tissues are resorbed and the adult rudiment develops.

Radial cleavage *Cleavage* in which the axis of cell division is directly perpendicular to the mother cell, so that cells lie directly beside or on top of one another (contrast *spiral cleavage*).

Radial symmetry Symmetry about any plane perpendicular to the oral/ *aboral* axis.

Radula Rasping tongue located in the *buccal cavity* of a mollusc: unique and diagnostic feature of the phylum. Most commonly a chitinous strip with teeth constantly renewed, but may become a drill or piercing organ.

18S rDNA Gene coding for *18S rRNA*.

Recessive Genes expressed only when present as both members of an *allelomorphic* pair; masked when the other member of the pair is different and *dominant*.

Recombination Rearrangement of genes occurring at *crossing over* in *meiosis*.

Regulative In embryological development, applied to late *commitment* of cells, as in most *radially cleaving* eggs.

Resilin Uniquely elastic *protein* forming elastic pads in arthropods, capable of storing and releasing large amounts of energy. Especially important in insects, to meet the demands of flight.

Respiratory pigment *Protein* molecule, pigmented and containing metal, that combines reversibly with oxygen and therefore can transport or store it.

Resting potential Small potential difference across a nerve membrane, the outside being positively charged, due to outward movement of potassium ions.

Retractor Muscle, contraction of which withdraws a structure back into the body.

Rhabdome Closely packed *microvilli* containing the visual pigment, e.g. in the central region of the retinula cells of an arthropod *ommatidium*.

Rhynchocoel Separate fluid-filled cavity dorsal to the gut of a nemertine worm, lined with mesoderm and containing the eversible *proboscis*.

Ribosomes Granular bodies in the cytoplasm of all cells, containing *RNA* of several kinds. Sites of assembly of amino acids, translation of the genetic code and *protein* synthesis.

RNA (ribonucleic acid) Nucleic acid made from ribose and other constituents of *DNA*. Differs from DNA in that the four bases are adenine, cytosine, and guanine, with uracil in place of thymidine. Various forms of RNA occur in all cells, often concerned with *protein* synthesis but with different functions including transcription of the genetic code from *DNA* and its translation at the *ribosomes*. RNA is itself the genetic material in some viruses, and perhaps was so at the origin of life.

18S rRNA The small subunit of ribosomal *RNA* (18S describes the centrifugation process isolating that subunit) often used in molecular taxonomy.

Rudimentary Structure at an early stage of its evolution, recognisable by comparison with its appearance later in evolution (contrast *vestigial*).

Schizocoel *Coelom* formed by splitting of embryonic mesoderm, as in most *protostomes* (contrast *enterocoel*).

Secretion Cellular manufacture of useful substances and the act of discharging these from the cell.

Sedentary Attached to a substratum but able to move away from the place of attachment (contrast *sessile*).

Segmentation Subdivision of an animal into a number of equivalent parts, contained between the anterior and posterior end of an organism (i.e. *metameric segmentation*).

Sensillum (pl. **sensilla**) Hairs, bristles or pads on the arthropod skeleton which are sensitive to mechanical or chemical changes in the environment.

Septum (pl. **septa**) Internal divisions within an animal (e.g. between the segments of an annelid worm).

Sequencing Of *genes*: identifying all the nucleotides in that part of a *DNA* molecule that constitutes a gene and finding their relative positions.

Sessile Organisms permanently attached to a substratum (contrast *sedentary*).

Seta See *chaeta*.

Sexual Reproduction involving the fusion of two *gametes*, each containing half the number of *chromosomes* characteristic of the species.

Shared derived characters Shared characters not *primitive* to the group concerned.

Shrimp Strictly applies to one group of decapod crustaceans but loosely applied more widely to any crustaceans filter-feeding with their swimming appendages.

Skeleton Structure providing mechanical support for the body and (usually) enabling muscles to be re-extended after contracting. May be hard or *hydrostatic*.

Somatic The part of a body that is not specialised for sexual reproduction.

Species (singular **species**, abbreviated sp.; plural also **species**, abbreviated spp.) Population of interbreeding organisms that normally produce fertile young. A species is designated binomially, by a generic followed by a specific name.

Spicule Small spine or spike, e.g. the hard structures in a sponge.

Spiracle Opening at the body surface, e.g. of a part of the *tracheal* apparatus of a terrestrial arthropod.

Spiral cleavage Early cleavage method where the axis of cell division at an oblique angle to the mother cell, so that the cells form an upward spiral as they divide (contrast *radial cleavage*).

Stenohaline Aquatic animals restricted to a narrow range of salinity.

Stomocord Structure in the collar region of hemichordates supporting the heart/excretory complex. Superficial resemblance to a short *notochord* is shown to be misleading by the contrast between the two in development and function.

Striated muscle Fast-contracting muscle with the *actin* and *myosin* filaments arranged to give a striped appearance under a light microscope (see Box 5.1). Occurs in nearly all animal phyla.

Stylet Hard pointed structure able to penetrate the tissues of other animals.

Success Survival defines success, for a living organism. The concept may include long persistence down the ages, large numbers of individuals (and of species, i.e. diversity), but not complexity or 'progress'.

Supraoesophageal ganglion *Brain* of arthropods and some other animals.

Suspension feeding Capture of food particles floating in water.

Symbiosis Close association of two organisms of different species that confers mutual benefit.

Synapomorphy Presence of *shared derived characters*.

Synapse The gap between two nerve cells or a nerve and a muscle or gland. Conduction across a synapse is mediated either chemically or electrically and occurs in one direction only.

Syncitium Group of cells without cell boundaries between the *nuclei*, whose division has not been accompanied by *cytoplasmic* division.

Tagma Region of an arthropod body containing a number of *segments* similar in size, shape and/or function.

Tanned protein *Protein* material held firmly together by many cross-links within the molecules. Hard, tough and dark, it provides the hardness of arthropod *cuticle*.

Taxonomy Systematic ordering of the products of classification of organisms.

Tegument Particular type of outer layer in place of the *epidermis* in parasitic platyhelminths.

Tissue An association of similar cells, usually connected by intercellular material. Organs consist of a number of coordinated tissues.

Tornaria Characteristic *larva* of enteropneust hemichordates.

Torsion Twisting of the body during development through 180 degrees, as is characteristic of gastropod molluscs. Torsion brings the mantle cavity from the back to the front of the animal, makes the gut U-shaped and may twist the nerves into a figure of eight.

Trace fossils Imprints on the environment left by animal activity of long ago in the form of trails, burrows or impressions of soft tissues.

Trachea (pl. **tracheae**) Air-filled cuticular tube, part of a system conducting the atmospheric environment to the tissues of a terrestrial arthropod (to every cell of the body, in insects).

Tracheole Fine terminal tip of an insect *trachea* penetrating between or even into cells. Cuticular, but not shed at moulting.

Triploblastic Having three cell layers, *ectoderm*, *mesoderm* and *endoderm*, as in nearly all animals.

Trochophore Early planktonic *larva* with two ciliary bands in front of the mouth and ciliary tufts elsewhere, as in annelids and molluscs.

Tube foot Thin-walled epidermal projection of the coelomic *water vascular system* in echinoderms. Unique to the phylum, tube feet are sensory and respiratory and often involved in feeding and locomotion.

Ultrafiltration Filtration on a small scale. Pressure (in most excretory organs) or suction (*flame cells*) forces water and some solutes through a semi-permeable membrane into a tubule.

Ultrastructure Level of structure that is almost molecular, being too small to be seen by light microscopy, but able to be studied by electron microscopy.

Uniramous Unbranched (i.e. having a single branch only).

Unsegmented See *segmentation*.

Urea $CO(NH_2)$, readily made by condensing two molecules of ammonia with one of carbon dioxide. The main nitrogenous excretory end point of many aquatic animals.

Uric acid Insoluble end point of nitrogen metabolism in many terrestrial animals, since it can be stored dry and excreted with very little water loss.

Vacuole Small membrane-bound intracellular sac containing fluid. May contain food, enzymes or water voided at intervals as in *contractile vacuoles*.

Variation Difference between individuals within a *species*.

Vegetal Yolk-containing region of an egg or *blastula*, at the opposite end from the *animal pole* where the nucleus is situated.

Ventral Lower or front surface, 'tummy'.

Vestigial Part of an organism that has diminished during its evolutionary history but not quite disappeared (contrast *rudimentary*).

Viviparity Developing young are housed in the mother's living body and fed from her own resources.

Water vascular system Transport system in which the transporting fluid is water, not blood, as in sea anemones and the *coelom* bearing the *tube feet* in echinoderms.

Zooid Individual within a colony; applied to Bryozoans and a few other animals.

Zygote Product of fusion of gametes in sexual reproduction, the single *diploid* cell from which an organism develops.

Index

Bold numbers denote figures and major accounts of groups or topics.